SUPPLIER MATTERS

Dr. Aditya Verma

INDIA · SINGAPORE · MALAYSIA

Notion Press

Old No. 38, New No. 6
McNichols Road, Chetpet
Chennai – 600 031

First Published by Notion Press 2019
Copyright © Aditya Verma 2019
All Rights Reserved.

ISBN 978-1-68466-489-4

This book is dedicated to

My mother, Late Smt. Maya Mayee,

Who overcame many hardships and dedicated

her life to caring for others

With an incredible amount of patience.

Loved by all, understood by few,

She spread love and compassion all through her life.

She was much ahead of her time in her thoughts and mindset.

She continues to inspire and guide us,

as we move on in our lives.

MATTERS!!!

In an increasingly competitive world, it is the quality of thinking and the simplicity of an application that give an edge to help stay ahead in the Procurement World. An idea that solves a problem or an insight that substitutes experience is the prime objective of this book for all those who are working in the area of procurement, sourcing, and supply-chain management.

Dr. Aditya Verma

Table of Contents

Preface

In today's highly competitive global marketplace, the pressure on organisations is to find new ways to create and deliver value to organisational growth. Gradually, in emerging economies as well as developed markets, the importance of the buyers has increased to stay competitive and relevant.

Rules and challenges are different in emerging and developed market and most of the times it varies based on industries, we are associated with. In such cases, procurement services and supplier development strategy becomes a key differentiator to drive supply-chain performance which eventually affects business.

Price has always been a critical competitive variable in global markets especially with respect to the chemical industry (majority of the case study in this book are in reference to the chemical industries in India, but are equally relevant to other industries as well with respect to supplier selection process) and the signs are that it will become even more of an issue as the commoditisation of the market continues, without sacrificing expectations on other fronts which are equally critical for sustainable relationship.

There has been a growing recognition that it is through "Robust Supplier Development" module, the goals of cost reduction and management and service enhancement can be achieved. Better management of the Supplier means nothing but creating a scenario where customers are served more effectively and yet makes the cost of offerings globally competitive.

The focus of this book is to provide first-hand insight on key attributes the buyer community should take care of while developing suppliers. There is an old saying, "Right supplier is key to delivering the right quality of product at right time in right quantity, at the right place and at the right cost with zero risks." The basic theme and structure of this book are that the best practice gets even better.

Price has always been a critical competitive variable in global markets, and, somewhere, because of that aspect developing, "RIGHT SUPPLIER" is getting lost. Many times organisations focus so much on price that letter "R" of SUPPLIER is either broken or demolished, to get RIGHT SUPPLIE (Supply), breaking the fundamental rule, that for right supply, we need to have a right supplier. Focused supplier development strategy would help in defining this "R" as Robust & Reliable. These signs and trend become even more of an issue as the commodification of market continues where a customer is looking for the best sustainable value proposition without sacrificing expectations on other fronts which are equally critical for sustainable competitiveness.

In a competitive environment, outsourcing is the mainstream practice in global business operations. In this context, vendor selection is a critical factor affecting outsourcing performance. Wrong vendor selection, is a key problem of companies offshoring procurement activity. Selecting and evaluating the right source is imperative for an organisation's global competitiveness. Improper selection and evolution of potential suppliers can dwarf an organisation's supply-chain performance. Therefore, material suppliers can play a very large role in the success of manufacturing companies. However, an increasing number of small and medium-sized enterprises (SME) are selected as strategic supply partners to global companies. This book also investigates Indian chemical companies' supplier selection process under three perspectives—Management-based criteria, Buyer's criteria and Customer perspective.

Given their importance, it is reasonable to expect the appropriate resources and methodologies that are employed in selecting the best suppliers. This book contends that for many firms, the reality does not match with their plans and execution. While firms may have vendor selection procedures in place, they are often ignored or are ineffective due to the complexities involved in the process or due to lack of education and training for execution and leadership commitment towards the programme. Either programme does not offer metrics important to the firm or may not include stakeholders within the firm such as quality, logistics, sales and marketing, operations and cross-function do not see value proposition out of that. An analysis of important criteria is done to simplify the vendor selection process through empirical survey within chemical industries in India, from small to medium to large size chemical companies.

One important aspect of supply-chain management is purchasing and selecting an appropriate supplier to ensure the success of the purchasing function. The importance of supplier selection increased with the wide implementation of JIT (Just-In-Time) among the manufacturing industry since the 1980s.

Literature findings estimated that 85% of North American and European multinational companies practice outsourcing. In 1966, Gary Dickson proposed 23 vendor selection criteria surveying 273 purchasing managers and agents, in a much-diversified industry set-up to rank them in order of importance. The study was enhanced by Weber in 1991, by reviewing 74 publications from 1966 to 1990. Zhang in 2004, compared Dickson and Weber's studies and summarised new supplier selection criteria from the study of 49 articles from 1992 to 2003. Along with the output of the above-mentioned studies, this book also covers a survey of 61 Indian chemical companies having a business turnover in the range of $100K to $50 Billion and above (small to large size firms), to reaffirm criteria used for supplier selection. The purpose is to identify top 10 new key attributes in Indian Scenario/w.r.t. chemical industries to be used for vendor selection. These have been evaluated and validated indicating that the cost, quality and delivery remain the greatest concern while selecting new suppliers.

Today when the market is characterised by globalisation, it has increased the customer's value expectations, expanding regulatory compliance, global economic crisis and intense competitive pressure. Therefore, to survive manufacturing operations, buyers must select and maintain core suppliers. Thus, supplier selection, evaluation and relationship management represent one of the significant roles of purchasing and supply-chain management functions.

Earlier days when supply-chain was not taken very seriously, the entire focus, initiatives and energy of an organisation were invested in "Sales and Marketing". But now it is a thing of the past. The supply-chain is now a significant contributor to the overall operational efficiency of the organisation, The small to large organisations are giving all required attention to supply-chain to stay ahead of the competition by leveraging global supply-chain or outsourcing methodology to low-cost countries to stay competitive.

Besides working for 16 years in Specialty Chemicals/Polymers Industries, the author has been associated with FMCG, i.e., Food/Personal Care/Health Care and OTC manufacturing global companies for over 10 years and hence tried to create a fusion of supplier selection process based on priorities of this key industry segments. Today when the whole world is looking for a cost competitive and qualitative supplier, to stay competitive in the global market, responsibility on procurement function has also gone up in multi-folds wherein, the right supplier selection can only deliver the growing buyer's expectations.

Thanks to Mr. Kenneth Gayer for suggesting that I write about the internal challenges and barriers based on my experiences in larger multinational companies, where buyers have to face unique complexities either in terms of getting everyone aligned or managing expectations contrary to the market situations. Many times, there are unwritten organisational structures or charts wherein everyone is a procurement expert over and above the procurement team and sometimes push procurement to the bottom of the pyramid.

This book gives a detailed outlook with respect to the sourcing aspects and deals with entire gamut of supplier development including source selection qualification programme and risk assessment to supplier governance module converting them into smart sourcing and addressing key perspective like Procurement, Supply-Chain Management, Supplier Relationship Management, Emerging Trends in Procurement Function, Negotiation & Cost Reduction strategy through Portfolio Management, Supplier Mentorship Programme and its direct impact on business to signify the title **"Supplier Matters."**

Sourcing Matters!

Global sourcing will be the next battleground for competition. Within the next few years, businesses that do not make global sourcing an integral part of their supply-chain process will be struggling not only to compete locally and internationally but also to survive in their business.

Let's face it; the days of your organisation are numbered if you are not focusing globally as a sourcing professional. Every sourcing professional knows that delivering continuous cost improvement requires a strategic approach to sourcing. It is very important that the sourcing initiatives and the related saving drives are not limited to paper only but also shows up on the bottom line of the organisation's cost structure, for which it is an important drive to make a low-cost supplier as a preferred source. It is equally important that the buyer must feel that ordering and receiving the right item from a preferred supplier is a painless process with a successful outcome, failing which global sourcing drive may result in frustration with an inefficient and ineffective buyer–Supplier Relationship.

Outsourcing as a term does not do justice to the flexible, relationship-based sourcing which requires open and partnership based method of delivering value to the organisation. It should be viewed as a partnership with third parties, which requires frequent assessments, adjustments and modifications. Primarily, sourcing can provide the potential to realise value, but it also has the potential to create value for an organisation. Strategic sourcing enables the organisation to focus on what they do best to leverage their key competencies and to create market space in doing so. Therefore, the identification and selection of partners are critical to strategic sourcing success. Sound partnerships are built upon reciprocity, commitment and shared vision by all engaged parties.

Every company wants their top line to grow without compromising on the bottom line. A strategy that ignores global competitive sourcing opportunities may cost a business both its customer and vast sums of money. An organisation needs a competent sourcing function to facilitate fair competition and to protect the company and its customers. Too often, however, sourcing has the opposite effect. Hindering competition and protecting some groups at the expense of a company happens when the company starts losing its profitability in the short term and customers in the long term. Poor sourcing competency is the main factor limiting productivity, profitability and growth of the organisation, particularly in companies where sourcing is country-centric, and they are not able to keep pace with fast-growing global competition.

For instance, sourcing chemicals are not like buying wheat or cement, etc. It is required to establish uniformity among the chemicals you purchase, use and store. When you buy the same chemical from five different companies, you end up buying five different levels of quality. All this adds up to the confusion for quality/process control functions for setting up different process standards in manufacturing and safety standards.

If we are serious about getting "Chemical Sourcing Act" together, we must look at more than just the price of the chemicals we buy. For example, what does a chemical accident cost in terms of money and lost time, cleanup costs, injury and in today's world where legal litigation costs are huge? We have many such examples in front of us in the chemical industry across the globe. Will a court of law have sympathy with our company when an accident occurs and they find out that we purchased a lesser quality chemical package because it was $3.00 per kg cheaper than the material we should have purchased?

Improving the overall safety profile of our supplier base as well as adopting stringent purchasing specifications for our chemicals/laboratory chemicals is the key to the success of competitive global sourcing function. Not all chemicals are equal in quality and packaging. Our production team can get properly packaged chemical consignments if the purchasing specifications are carefully written by keeping in mind the safety requirements for handling the consignment at the testing laboratory and plant level, yet at a competitive cost.

Suppliers Matters as the driving forces for companies to offshore their businesses and functions are globally integrated by labour and capital markets that are facilitated by better technology. Behind these developments, there are factors like improved infrastructure, communication ease and friendly business environment. A procurement or sourcing strategy is a strategy in which a business seeks to find the most cost-efficient supplier location for manufacturing a product, even if the location is foreign.

For example, if a chemical manufacturer finds that manufacturing and delivery costs are low in a foreign country due to the low wages of the foreign employees, the company may close the domestic factory and use a foreign manufacturer. In sequence, the former competitive landscape, mainly dominated by Multinational Companies (MNCs) are facing new competitors today. International markets enable the participation of companies in any size, which resulted in an increased number of globally spread small–medium enterprises. With this business objective, the major task that comes for any procurement wing is the selection of the right suppliers to meet organisational objectives with respect to cost, quality, delivery and performance.

The contracting or subcontracting of non-core activities to free up cash, personnel, time and facilities for activities in which a company holds a competitive advantage. Companies with strengths in other areas may contract out some aspects of their business to concentrate on what they do best and thus reduce average unit cost. Outsourcing is often an integral part of downsizing or re-engineering to support core business objectives and to identify the most cost-effective supply solution or product manufacturer across the globe.

Outsourcing as a term does not do justice to the flexible relationship-based sourcing which is an open and partnership based method of delivering value to the organisation. It should be viewed as a partnership with third parties, which requires frequent assessments, adjustments and modifications. Primarily, sourcing can provide the potential to realise value, but it also has the potential to create value for an organisation. Strategic sourcing enables the organisation to focus on what they do best to leverage their key competencies and to create market space by doing so. Therefore, the identification and selection of partners are critical for successful strategic sourcing. Sound partnerships are built upon reciprocity, commitment and shared vision by all engaged parties.

Choosing the right supplier model or configuration of attributes for necessary screening is the essential first step. Customers should assess a supplier's capabilities and competencies for each new business context, because not every business context requires suppliers to excel in all capabilities/competencies/attributes selected for supplier screening.

In assessing suppliers, there are 5 major criteria such as Qualitative, Commercials, Logistics, Organisation and Infrastructure. If the right selection skill set is not used at this stage, there can be adverse repercussions. The vendors need to fit the sourcing company's culture and meet technical requirements in order to ensure mandatory and consistent quality. Obvious and critical aspects during the evaluation phase include price levels, lead time, quality and technological capabilities.

The official foundation of the European Community (EC) 1992 under the auspices of the single European Act, the demise of communism in the eastern bloc countries and the pending ratification of the North American Free Trade Agreement contributed heavily to the globalisation of the world economy. The progressive lowering of trade barriers, the advancement of information and communication technologies, the development of transport systems and infrastructures have facilitated international trade and have increased the level of competition worldwide.

Intensifying degrees of global competition today greatly accelerated the growth in international sourcing. It allowed firms to utilise worldwide resources more effectively, by enabling them to decouple regional economies from their countries of origin. It enhanced purchasing from suppliers located outside the national border making the "Supplier Selection Process" critical, important and strategic due to presence of factors such as complicated documentation requirements, trade regulations, quotas, custom duties, currency exchange rates, cultural differences, unique ethical standards in addition to complex distribution channels which made it a more complex decision process compared to the domestic sourcing. Eventually, many buyers unfamiliar with these factors often hesitate to internationalise their sourcing processes.

Supplier selection problems usually consist of multiple criteria that contradict each other. However multiple criteria decision-making (MCDM) analyses assume that these criteria are independent on each other. Internationalisation of sourcing requires the selection of suppliers in the international arena while considering political, social, economic and environmental dimensions in related countries.

The goal is to select the best supplier. Weber (1991) attests, "It is impossible to successfully produce low-cost, high-quality products without the selection and maintenance of a competent group of suppliers." A number of methodologies that has been used in supplier selection and evaluation studies include linear weighing models, the categorical model, weighted point model, total cost of ownership, multiple attribute utility theory, artificial neural network, principal component analysis, analytic network process, analytic hierarchy process (AHP), combined AHP – A linear programming, etc.

The most successful companies have configured their supply-chains for specific customer segments, adopted differentiating practices such as collaborative planning with customers and suppliers and reduced complexity. Next generation supply-chain is fast, flexible and responsive—a model that enables companies to serve their customers accurately and efficiently in turbulent market conditions and differentiates between the need for different sets of customers.

Companies which focus on improving their supply-chain performance achieve much better financial and operational results than their peers do. Even supply-chain executives themselves usually don't realise the full value they bring to their organisations. That's because most supply-chain executives are focusing on the day-to-day aspects of establishing and managing an end-to-end supply-chain fostering collaboration, both with other functions and within the supply-chain itself. But taking the time to promote the importance of the sourcing and supplier selection within the supply-chain function can have significant benefits, as it is vital to meet the requirements of customers, who are becoming more demanding about the delivery performance, flexibility and service level they expect from supplier base.

The supply-chain also needs to support demand growth in emerging regions/markets and be more sustainable. Most companies have, so far, devoted relatively little effort to the idea of the sustainable supply-chain, largely because their customers seem unwilling to pay for it. But the importance is now rising sharply and it will play a major role in the next five to ten years of time compared to now.

Offshore outsourcing has become an international mainstream activity which largely determines a company's success level. Consequently, the success level is highly dependent on the sourcing company's ability to select appropriate vendors. Offshore outsourcing deals with two concepts including geographical and legal aspects. The geographical concept mainly means to submit relocation of a value chains further than national borders. The legal element, in this case, is subcontracting.

SOURCING IS AN ART AND SCIENCE

- Sourcing is an art of spending in order to achieve the desired objectives as well as save money in the process. There is a saying that a single dollar saved in procurement function adds almost 100% to business profitability. There are three options for sourcing:

  - It has to be operational and highly efficient.

  - It has to be professional.

  - It has to be strategic to create value.

The word "Sourcing" is derived from the word "Source", which is one of the most important elements of the purchasing function. The other elements of purchasing like quality, quantity, time, price and place all are sub-part of the source. Therefore, right source selection is critical to all sourcing deliverables.

Sourcing is science in terms of right strategy formation which addresses innovations to improve performance and leads to customer delight by offering quality on time, by challenging the procedures to improve the results. Purchasing strategy formation takes full account of supplier innovation, supplier involvement and innovation protection to deliver sustainable, beneficial results to the organisation.

WHAT TO EXPECT FROM PURCHASING

The purchasing process can be defined in three phases, i.e., Need phase, Set-up phase and Execution phase. During the Need phase, any purchasing personnel is expected to deliver on following,

- Identify the right need

- Avoid monopoly or sole-source

- Identify potential suppliers

- Secure competitiveness

- Consolidate volumes

- Use of frame agreements

- Manage supplier integration

- Identify and cover risks
- Show the economic impact of technical discussion

Whereas during the Execution phase, they need to deliver following—

- Get the best deal from the market
- Meet cost reduction targets
- Deal with the price increase
- Steer the supplier selection process
- Use cost analysis break down for negotiation
- Analyse quotation
- Reduce switching cost
- Monitor supplier performance
- Deal with shortages and arrange on-time supplies
- Deal with compliance related issues
- Co-operative planning
- Manage supplier development
- Establish an integrated relationship
- Deal internally towards resistance to change
- Work in reactive mode

Normally, it is seen that a procurement person spends 75% of his time handling issues related with logistics and administration and 25% time in core purchasing activities and hardly any time on a day-to-day basis for strategic sourcing related activities. Whereas, ideally a sourcing guy should ensure that they spend 60% of their time for core purchasing activities which includes cost saving, supplier base review, manage supplier base, procurement planning, etc. They spend 20% time for strategic sourcing activities like purchasing process improvement, strategic make or buy decision, how to manage portfolio, supplier market analysis, etc. They spend the leftover 20% time for other logistics and administrative activities. Any imbalance in this distribution may directly affect the quality of output and efficiency of the function.

MAJOR CHANGES IN PURCHASING

Like other things, Purchasing has also changed a lot in the last few years. Here are the top 10 purchasing changes in last few years.

1. **Technology Proliferated** – Today, e-Procurement/e-Sourcing and e-Auction are two of the most useful practices in purchasing. Ten years ago, these terms were unheard.

2. **Centre-Led Procurement Arrived** – In 1998, even the top purchasing departments processed the purchase orders. Today, purchasing departments aim to centralise the supplier selection process, not the transactions which are delegated to end users or outsourced.

3. **Purchasing Grabbed More Spend** – When purchasing departments deliver results, management spends more that can impact Purchasing positively. Once sourced by other departments, categories like fleet management, benefits, travel services, MRO, consumables and refreshment expenditures are now facilitated by Purchasing.

4. **Social Responsibility Became a Top Priority** – Whether for philanthropy or to avoid media scandals, management counts on Purchasing more than ever to buy from diverse suppliers, making environmentally-conscious decisions and doing business ethically.

5. **Measurement Was Mandated** – With the potential of smart purchasing widely known, senior management is more strict withholding their purchasing departments accountable for results. The use of purchasing metrics and dashboards is now commonplace.

6. **Strategic Sourcing Becomes In-House** – In the 90s, strategic sourcing was done mostly by consulting firms hired to help companies reduce their expenditure. Today, many companies have their own refined and documented in-house strategic sourcing processes.

7. **Supplier Roles Expanded** – In 1998, there was talk about partnering with the suppliers. Today, there's an action. Top purchasing departments actively develop their suppliers and look to the supply base for ideas, better performance and innovation.

8. **Global Sourcing Is Now Mainstream** – Ten years ago, only the progressive companies were searching abroad for suppliers. Now in some countries, it is difficult to find products manufactured domestically.

9. **The CPO Position Got Adopted** – The last few years alone, we've encountered an unprecedented number of folks with the title "Chief Procurement Officer." Now irrespective of the size of the company, be it medium or large scale, this position is an integral part of senior management team.

10. **Procurement Converting into Supply-Chain** – In the last decade, companies more closely analysed the way material flows into, through and out of the organisation. Now it's not mere procurement, it's "supply-chain" where the focus has changed from those who once just placed orders are now responsible for inventory, warehousing, outbound logistics and distribution.

WHY MARKET INTELLIGENCE IS CRITICAL FOR PURCHASING

"**Market Intelligence** is the process of gathering and analysing information relevant to a company's supply market specifically for the purpose of supporting accurate and confident decision-making in the procurement process."

- **Commodity** *Profile* – This lists the information about the products/services classification, market size and growth rate, manufacturing process and critical-to-quality factors.

- **Cost** *Structure* – This section identifies the costs associated with materials, labour, transportation, energy, overhead, profit and other cost components.

- **Supply-***Base Profile* – This identifies available suppliers, their characteristics and the countries where suppliers of the category are located.

- **Market** *Indicators* – This identifies demand and price drivers, capacity utilisation and other characteristics that determine price and availability.

- **Competitive *Analysis*** – This assesses buyer and supplier power, substitute products/services and other factors that influence your buying leverage.

Conducting market intelligence can reduce risk, increase savings, improve decision-making and offer the ability to challenge some of the assumptions you've had. If you regularly bid out a given commodity or spend area to the same supply base, you are always going to get similar results.

EXCHANGE RATE RISK IN INTERNATIONAL PROCUREMENT

When buying from a foreign supplier, one of the critical things we need to negotiate is which currency will be used for pricing and payment. Why is this critical?

Well, imagine that you are a US buyer, buying GEARS from a supplier in India. You agree to purchase 1 million GEARS for 600,000,000 Indian Rupees.

On the date of your agreement, i.e., January 1st, the exchange rate is such that 600,000,000 India Rupees equals 10,000,000 US Dollars. You are required to pay up when the shipment arrives. On the date of shipment, i.e., April 1, the exchange rate is such that 600,000,000 India Rupees now equals 11,000,000 US Dollars.

Your company just paid 10% more than it expected to. You don't have that type of challenge with domestic purchasing.

So how do seasoned international buyers protect their organisations from this type of risk?

One way is by acquiring forward contracts. A forward contract is a method of locking up an exchange rate today for a transaction in the future. So in this situation, if you had acquired, on January 1st, a forward contract to receive 600,000,000 India Rupees in exchange for 10,000,000 US Dollars on April 1st, you would not have endured that 10% loss.

So while buying internationally, remember these things:

- It is important to decide carefully the currency of payment.

- Your own currency may not be the best currency for the purchase.

- If buying in the seller's currency, consider acquiring a forward contract to mitigate the risk of an exchange rate fluctuation.

UNDERSTANDING DYNAMICS – PORTFOLIO ANALYSIS

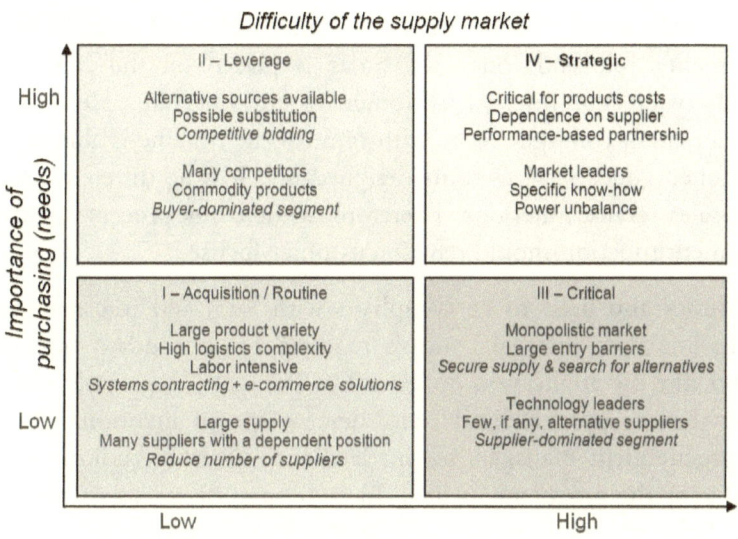

Difficulty of the supply market

	II – Leverage	IV – Strategic
High	Alternative sources available Possible substitution *Competitive bidding* Many competitors Commodity products *Buyer-dominated segment*	Critical for products costs Dependence on supplier Performance-based partnership Market leaders Specific know-how Power unbalance
Low	I – Acquisition / Routine Large product variety High logistics complexity Labor intensive *Systems contracting + e-commerce solutions* Large supply Many suppliers with a dependent position *Reduce number of suppliers*	III – Critical Monopolistic market Large entry barriers *Secure supply & search for alternatives* Technology leaders Few, if any, alternative suppliers *Supplier-dominated segment*

Importance of purchasing (needs) — Low / High

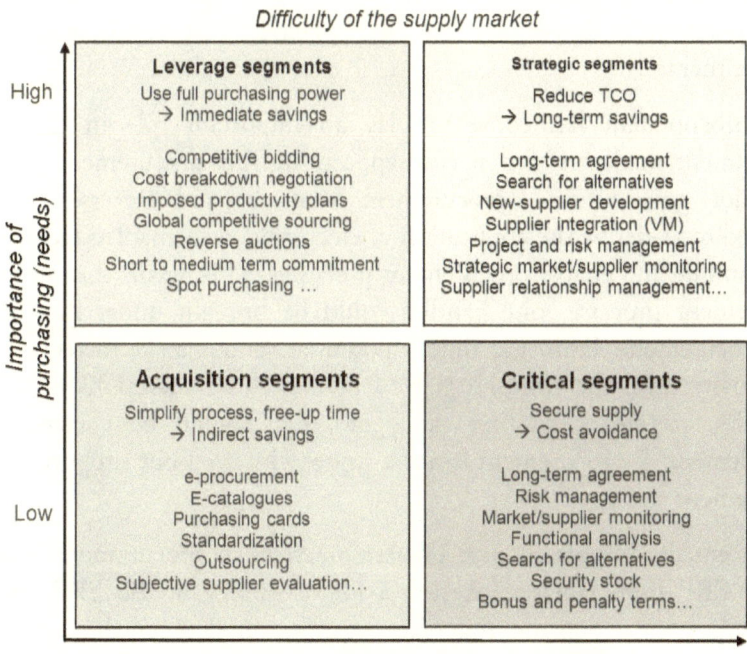

Difficulty of the supply market

	Leverage segments	**Strategic segments**
High	Use full purchasing power → Immediate savings Competitive bidding Cost breakdown negotiation Imposed productivity plans Global competitive sourcing Reverse auctions Short to medium term commitment Spot purchasing …	Reduce TCO → Long-term savings Long-term agreement Search for alternatives New-supplier development Supplier integration (VM) Project and risk management Strategic market/supplier monitoring Supplier relationship management…
Low	**Acquisition segments** Simplify process, free-up time → Indirect savings e-procurement E-catalogues Purchasing cards Standardization Outsourcing Subjective supplier evaluation…	**Critical segments** Secure supply → Cost avoidance Long-term agreement Risk management Market/supplier monitoring Functional analysis Search for alternatives Security stock Bonus and penalty terms…

Importance of purchasing (needs) — Low / High

SIMPLIFY YOUR PROCUREMENT PROCESS

Procurement processes need to be simple and friendly to the internal customer. One should find the procurement department easily approachable. The processes must have the shortest turnaround cycle from the time a requisition duly approved is received, to the time requisitioned goods/services arrive. There are many who would like to believe that procurement's role ends once the order is placed on the supplier. This is exactly what the internal customer would not like. He would like the procurement colleague to be with him till the time he is able to put the requisitioned goods/services to its designed use. It is no different from sales or after-sales service functions. Therefore, while being process-oriented, the procurement function should remain customer-focused.

Processes also need to be compliant with local and international laws to be in line with internal controls required for a standard organisation. Concepts like the application of "Four Eye Principle", segregation of duties, hierarchy-based approval limits, etc. deserve special mention here. Quite often, procurement managers see internal and statutory auditors as people who go after the procurement jobs. In fact, an audit needs to be seen in a positive light and auditors should play the role of constructive critics who can help improve internal processes. Timely implementation of audit recommendations is yet another best practice that organisations must commit themselves to.

E-procurement has come to play an important role in the life of procurement managers. But it is disappointing that procurement managers have not got to use e-procurement tools beyond Reverse Auctions. We need to re-define the application of electronic, web-enabled technologies to streamline and enable procurement processes. This means that all routine procurement processes one handles could be brought under the purview of e-procurement. From the time a potential vendor seeks to register with a company until the time he gets to manage his transactions of delivery schedules, payments, posting catalogues, etc. should be considered for e-enablement. E-procurement is not a strategy by itself but supports a larger procurement strategy.

An equally important part of best practices in procurement is played by the ERP tool. Having invested a lot of money in the ERP package, it is important to ensure that the organisation deploys the application effectively. A study conducted a couple of years ago suggested that organisations do not use more than 15–20% of the functionality and feature ERP packages offers to the procurement professionals. It is, therefore, essential that organisations evaluate all options before deciding on one or the other application.

Further, the best practice that will stand by any procurement organisation is the way it manages its relationship with suppliers. Suppliers must get to have a fair say in the materials they supply or the services they provide since they possess the required core competencies in their field and must be respected for it, apart from being guaranteed fair returns on their investment and timely payments. Every supplier should be encouraged to innovate in order to create value for his customer continuously.

In a survey conducted with senior procurement professional, following five key areas have been identified which the procurement organisations are expected to address in a synchronised manner to stay ahead of the competition:

- Becoming business partners, not just buyers.
- Exploring new value frontiers, it is not just about price.
- Pulling suppliers inside, the best value chain wins.
- Pursuing low-cost sources, a world worth exploring.
- Conducting the ultimate talent search, do so in record time.

Adopting best practices will only put procurement professionals in a good position as this function moves from the back room to the boardroom. It will also bring a significant efficiency to the supply-chain, and quick adopters will surely enjoy a competitive advantage.

Finally, the incumbent would need to measure and forecast, analyse expenditure to establish potential future price to change impact on purchasing portfolio and the total forecast impact/purchasing activity on future business profitability.

New Supplier Selection Stages and Criteria

It is important to ensure that the supplier selection process in the sourcing function of any organisation addresses all involved risk and adequate provisions or processes which are in place to safeguard the buyer's long-term business interests. For instance, detailed working on overall storage and logistics-related cost involved with the supply-chain cost while comparing the sourcing cost from local source versus overseas supplier, product price benefit versus increase in inventory cost and increase in rejection risk, increase in lead time involved in the procurement process as well as substantial increase in overhead cost associated with additional infrastructure, people and other indirect cost involved with running these global sourcing offices outside the country.

- Identification of suitable Small–Mid–Large size companies into chemical business for comparative study and analysis. Based on sales turnover, number of employees, number of new products or new customers introduced every year. One must analyse organisations on the following along with their various sub-categories;

  ○ Company Goodwill/Market Reputation

  ○ People & Organisation

  ○ Delivery and Logistics

  ○ Supplier Quality Management

  ○ Financial Flexibility

  ○ Manufacturing and Innovation

  ○ Corporate HS+E and Social Responsibility

- Comparative analysis in tabulated form for various organisations based on the major attributes of source selection;

  ○ Management perspective

  ○ Buyer's perspective

  ○ Customer's perspective

- Aggregation of the score on various attributes based on survey data

- Identification of most popular/critical areas where 100% focus is required

- Recommendations
- If there is any related future extension of the project study

The scope of such studies is to compare the supplier selection process and criteria across and within companies using both qualitative and quantitative approaches. The focus is limited to companies operating in the Indian market with the global customer base.

Supplier Selection – Key Strategic Step

CHOOSING RIGHT SUPPLY PARTNER

It is becoming equally important for a foreign or local buyer to evolve a process of source selection in context to sourcing which has provision to deal with intense global competition, sustainability of the relationship, thrust for continuous cost improvement, total quality excellence in order to make this new association a win-win equation for seller as well as for buyer and ensure "Right Source" technology. It is almost desired of each and every company to deliver "Customer Delight", i.e., efficiency and effectiveness and hence what customer wants over and above the product becomes an important aspect. This also focus on the behavioural and cultural aspect which affects the business relationship.

Over the past few decades, new manufacturing concepts and methodologies such as TQM, JIT and ERP have irreversibly changed the role of the supplier. It has a threefold objective:

- To gain an insight into the level of seriousness while selecting a new source for supply of products by small–mid–large companies w.r.t. quality, cost, health, safety and environment and CSR, etc.

- To understand the buyer perspective, when they look for the right source or supply partner to support their global operations.

- To enumerate factors affecting supplier selection criteria in addition to various behavioural and cultural aspects.

The strategy of many organisations has become "Do what you do best", focusing on the organisation's core competencies and outsourcing operations that are not within their core competencies to external companies or suppliers. As a result of this trend, manufacturers have come to look at their suppliers increasingly to take on non-core manufacturing activities. Most Japanese companies in the automotive industry have proved the value proposition of a strong strategic relationship with their suppliers and have gained efficiency and productivity through their supply-chain. Therefore, the key objectives to satisfy are as follows:

i.   Criteria to identify potential new suppliers to stay competitive to survive.

ii. Identify the top 10 important criteria for supplier selection.

iii. Attributes that affect and drive supplier selection process to stay competitive, reduce supply disruption risk.

The objective of this research is to answer 1 (ONE) principle question, i.e., which are the major criteria of source selection in the organisation from three perspectives.

a. **From Management Perspective** – Key focus areas for supplier selection.

b. **From Procurement Perspective** – Key focus areas for new supplier selection.

c. **From Customer Perspective** – What makes a customer select a particular supplier?

Senior management must be involved in selecting and managing the supplier because while some organisations see outsourcing as an opportunity to pass on risk, in practice, such risk displacement is largely illusory. It is illusory because an organisation ultimately suffers if it chooses the wrong supplier.

On the other side, it is equally important that if other functions like Quality, Production and Business teams are also involved along with procurement in the assessment process of key or strategic supplies. This will enable new business association or supplier relationship to be strong, qualitative, sustainable and mutually beneficial.

Following are the broad objectives that can be achieved through this;

1. Determine an evaluation approach.

2. Develop a method to collect information about suppliers.

3. Identify the key selection criteria.

4. Design and develop an assessment system.

5. Address organisation specific needs and attach required weight to each criterion.

6. Deploy a supplier performance assessment system.

7. Produce result from measuring supplier's performance assessment programme.

8. Based on survey findings define, construct and propose assessment template for new supplier selection to fulfil major requirements in the industry with respect to quality, cost, delivery, sustainability, global competitiveness, etc.

9. Identify top selection criteria in a specific industry in India to meet global customer's expectations.

10. Explore if there is any future work required based on the outcome of the survey.

The major objective in supplier selection process or focus is predominantly assessing supplier capability to handle issues such as product availability, just-in-time delivery, cost and quality, end-to-end cycle time, whereas the buyers are increasingly finding sustainability challenges to introduce innovations to find new efficiencies.

Today's new economic environment is increasingly more volatile, complex and structurally different than in years past, this has underlined the most significant challenges supply-chain executive might face:

- **Volatility:** Fluctuation in customer demand has been the leading challenge confronted by supply-chain executives. Added to demand variances are increasing customer requirements for sustainable products and services as well as heightened expectations from responsiveness, uncompromising quality and low cost. At the same time, buyers are encountering poor quality and reliability performance from suppliers, which along with logistics constraints and bottlenecks, hampers delivery performance and customer service levels. As more companies continue to globalise their operations and enter emerging markets, these issues may have no quick solution, as operations will become increasingly dependent upon a growing number of customers, suppliers, regulators and markets.

- **Visibility:** As the number of supply-chain partners increase, the need for accurate, time-sensitive information becomes more acute. But lack of collaboration and integration between supply-chain and product development partners continues to be a major concern. Product lifecycle traceability in consumer products, chemicals and other industries is a growing requirement. Yet, despite continued technological enhancement, lack of visibility to worldwide, timely information to make in-stream decisions remains a significant issue. The bottom line is that the requirements for increased visibility require the dexterity to make fast decisions in response to constantly changing market conditions.

- **Value:** There is constant pressure for supply-chain management and operations to create enterprise value. End-to-end supply-chain cost and pipeline inventory optimisation are predominant challenges, as well as the means for protecting margin and decreasing working capital. Securing and deploying the right talent and skills for global operations remains a critical concern. The talent vacuum is most acutely felt in emerging markets, with a majority of executives citing it as a challenge. The business risks associated with insufficient leadership talent is exposed in decreased cost efficiencies, inventory deployment and in managing regional and local operations with partners. Managing risks and disruptions with global partners at each mode is increasingly important. As supply-chains become more complex and interdependent, managers must find a way to offset growing complexity with increased flexibility.

In short to choose suppliers, remember the following:

- Thinking strategically when selecting suppliers.
- What should be looked at while selecting suppliers?
- Identifying potential suppliers.
- Drawing up a shortlist of suppliers.
- Getting the right supplier for the business.

Why Is the "Right Source" Important?

In most manufacturing organisations, the role of purchasing and materials management has gained visibility and additional responsibility. It has gained recognition from top management as a key process that contributes to sustainable competitive advantage. The importance of source selection continues to expand as an organisation enhances the amount of outsourcing in order to focus on high-end processes. For instance, in the chemical industry, these efforts cause organisations to rely more heavily on their suppliers for the development and production of certain molecules and intermediates. As this alliance grows, performance increasingly depends on the quality of suppliers. The organisation continues to seek performance improvement. Therefore, this task of supplier selection has become strategic in nature.

To avoid the dire outcomes of supplier non-performance, buyers typically take proactive steps to verify a supplier's qualifications prior to awarding them with a contract. The primary purpose is to ensure and reduce the likelihood of supplier non-performance such as late delivery, non-delivery or delivery of non-conforming materials. Whereas, the larger goal is to ensure that the supplier will be a responsible and responsive partner in the day-to-day business relationship with the buyer. This aspect may include various screening methodology such as reference check, financial status check, surge capacity availability, an indication of supplier quality, ability to meet required specifications, buy-in from other internal customers, etc.

Supplier Selection Process reviews the diversity of procurement situations in terms of complexity and importance into account and covers all phases in the supplier selection process from initial problem definition to the formulation of criteria the selection of suppliers for qualification process and the final choice among the qualified suppliers. As this leads the organisation to become more dependent on suppliers and the direct and indirect consequences of poor decision-making become more severe, this aspect requires careful study and brainstorming to ensure sustainability. Also, changing customer preferences require a broader and faster supplier selection. Thus, the supplier selection process:

- Improves the efficiency of purchasing (management) decision-making.

  o Enabling more efficient storage of purchasing decision-making process and access to information in future cases, e.g., saving files that contain criteria-structures for supplier evaluation.

○ Eliminating redundant criteria and alternatives from the decision or evaluation process, e.g., in extensive and expensive supplier audit programmes.

○ Facilitating more efficient communication about and justification of the outcome of decision-making processes, e.g., when reporting to management or suppliers.

• Enhances the effectiveness of purchasing decisions by

○ Aiding the purchaser in solving the "right problem", e.g., refraining from dropping a supplier because of any outdated information.

○ Aiding the purchaser in taking more and relevant alternatives criteria into account when making purchasing (management) decisions, i.e., supplier selection.

○ Aiding the purchaser to model the decision situation more precisely, e.g., dealing specifically with intangible factors and group decision-making.

Thus, the goal is to evaluate and select the best suppliers for right alignment as per desired industry standards to cater to large customer base locally or globally.

Going forward, product lifecycle analysis becomes more important, which enhances a need to go for deeper long-term collaboration with suppliers to help them understand and invest in sustainability issues.

CHOOSE YOUR SUPPLIER SELECTION METHODS

Methodology	Reference	Parameters	Advantages	Disadvantages
Categorical	Timmerman (1986)	Quality Delivery Service Price	The evaluation process is clear and systematic. Requires minimum performance data.	Attributes are weighted equally. Subjective Imprecise

Methodology	Reference	Parameters	Advantages	Disadvantages
Weighted Point	Timmerman (1986)	Quality Delivery Service Price	Attributes are weighted by importance.	Difficult to effectively consider qualitative data.
Cost Ratio	Timmerman (1986)	Quality Delivery Service Price	Flexibility	Complexity Performance measures are artificially expressed.
Total Cost of Ownership	Ellram (1995)	Price Quality Cost Unreliable Delivery Transportation cost Ordering cost Reception cost Inspection cost	Substantial cost savings. Allow various purchasing policies to be compared with one another.	Complex
Principal Component Analysis	Petroni & Braglia (2000)	Price Delivery reliability Quality	Consider multiple inputs and outputs.	Knowledge of advanced statistical methods is required.
Analytical Hierarchical Process	Nydick & Hill (1992)	Quality Price Delivery Service	Simplicity Captures both qualitative and quantitative criteria.	Inconsistency on the method.
Neutral Network	Siying Wei (1997)	Performance Quality Geography Price	Saves a lot of time and money for system development.	Lack of expertise. Requires a software.

In conclusion, these methods are apparently useful for the supplier selection process. However, every industry may have its own priorities with respect to important or critical attributes.

In general, the three most important criteria are price, delivery and quality. These criteria were rated with extreme importance by Dickson and ranked top 3 by Weber and Zhang. These criteria are also known as order qualifier as this is the basic requirement for most purchasers. Other selection criteria namely, services, supplier relationship and management and organisation status are known as order winner. The buyers prefer the supplier who can provide better customer service, easy assessable technical support and consistent quality. These are the elements of good customer service. Relationships with the suppliers become increasingly important in the manufacturing industry. Good suppliers help manufacturers during the development of new products and process, with long-term quality improvement, cost reduction and enhance delivery performance.

In order to obtain a competitive advantage, companies are streamlining the numbers of suppliers. Reduced supplier base means a closer, long-term relationship that can be established with a few suppliers. A supplier is not only a seller but also a business partner for the purchaser's company. Geographical locations, financial status, performance record, company background and even culture can be important criteria which might affect the reputation of the supplier.

Globalisation has taken supply-chain into new emerging markets, making supplier selection as a critical aspect. Preparing for this challenge has required a wide-ranging supply-chain approach, encompassing;

- **Systems:** A robust quarantine management system to track every item throughout the supply-chain.

- **Processes:** Detailed management plans and directions for every aspect of the project, including specific requirements incorporated into related supply-chain contracts.

- **Compliance Monitoring:** To identify specific things potentially in breach.

- **Skills:** Developing new expertise within supply-chain from quarantine issues through to environmental assessments.

- **Collaborations:** Relationships with specialist partners.

- **Organisational Behaviour:** Induction on safety and environmental concern to remind workers of specific requirements as a fundamental part of the working process.

The supply-chain is playing a more central role in coping with the ongoing volatility and looming scarcity risks while pushing harder to operationalise competitiveness. This helps in giving the function a new dimension and depth, eliminating inefficiencies, improving competitiveness and differentiating the company's image. Over time, there will be a shift from trying to make existing supply-chains more sustainable towards developing wholly new ones that demand alternative business models. Little of this can be done in isolation.

At one extreme, marketing is needed for crucial customer insights; while at the other extreme, key decisions need to be closely coordinated with procurement, information technology, suppliers and other stakeholders. The secret of designing a superior supply-chain like build-to-order strategies is to start by re-segmenting customers along behavioural lines and then reverse engineering from there.

The whole area of supply-side sourcing is coming back into focus once again as the world recovers from the global financial crisis and seeks to reduce its impact on the real economy. Since the turn of the new millennium, multinational corporations, in particular, have been pursuing global sourcing strategies in the relentless search for ever lower cost inputs to manufacturing.

Most companies are still largely basing their procurement decisions on a minimum price approach rather than the more sophisticated total cost of ownership concept. Further, global trade appears to be significantly contributing to the emission of greenhouse gases because of the added transportation sectors involved, and this shows up in the face of the growing global efforts to reduce carbon dioxide emissions. Maybe changing back to local or green sourcing would satisfy the community's growing concern about the impact of climate change.

There is a clear trend emerging which is empathetic towards sustaining the natural environment and, consequently, is demanding corporate social responsibility. This sub-segment will surely penalise suppliers along the supply-chain who do not take sufficient measures to minimise their carbon footprint. The task ahead is to reconnect the supply-side to supplier selection. It is hard to imagine how an organisation can successfully procure the raw materials, packaging and other inputs for its business if there is not a live connection with the customer-facing side of the business, but that is what has been going on for generations and sadly continues to this day in many companies.

Suppliers are the largely ignored element in the human system that propels contemporary supply-chains, along with customers at the front end and employees inside the business. Companies that give time to listen to their suppliers can eventually realise rewards beyond cost savings. Often, they can generate new ideas and concepts, determine how their performance stacks up against their best practice competitors, and identify areas and processes on which to focus attention.

Major megatrends in sourcing/procurement functions emerging are as follows:

- **Outsourcing:** Supplier performs manufacturing tasks, which were formerly performed in-house.

- **Global Sourcing**: Companies moving purchase from domestic to foreign low-cost country sources in order to lower their costs.

- **Supply-Chain Optimisation:** Companies seeking suppliers to achieve "Build-to-Order" capability to reduce inventory carrying cost, obsolescence, spoilage and overstocks.

- **Supplier Consolidation:** In order to gain volume purchasing power and reduce administrative and coordination costs, companies have been striving to consolidate to fewer suppliers.

A common thread in the above approaches to improve purchasing is the supplier selection decision problem. Any mistake in this decision can easily render the approach ineffective and could even adversely affect the stability of the organisation in a current turbulent competitive environment.

UNDERSTANDING THE IMPORTANCE OF SUPPLIER SELECTION

The set of procedures is used to identify products for purchase, to verify quality and compliance of products and vendors, to carry out purchasing transactions and to ascertain that operations associated with purchasing have been executed appropriately.

Different organisations have been buying processes of varying complexity, depending on the industry in which they work and the nature of the products purchased. The purchasing process has been viewed as a set of operational activities incapable of contributing to overall organisational effectiveness.

In recent years, this view has changed, and today, considerable attention is given to the strategic contribution that a purchasing process can make to an organisation's Total Performance.

One should look into the strategy that should be pursued for raw material procurement by the Indian companies to operate it in a global environment. For instance, in case of chemicals, it is an important criterion that the chemical is handled appropriately and under appropriate conditions during the supply process and the purity and the impurity levels of demanded compounds are of great importance for buyers.

Next is the base framework for "Supplier Selection Process" in any industry:

- Develop the Survey

- Supplier Audit and Selection

- Continuous Supplier Performance Review mechanism

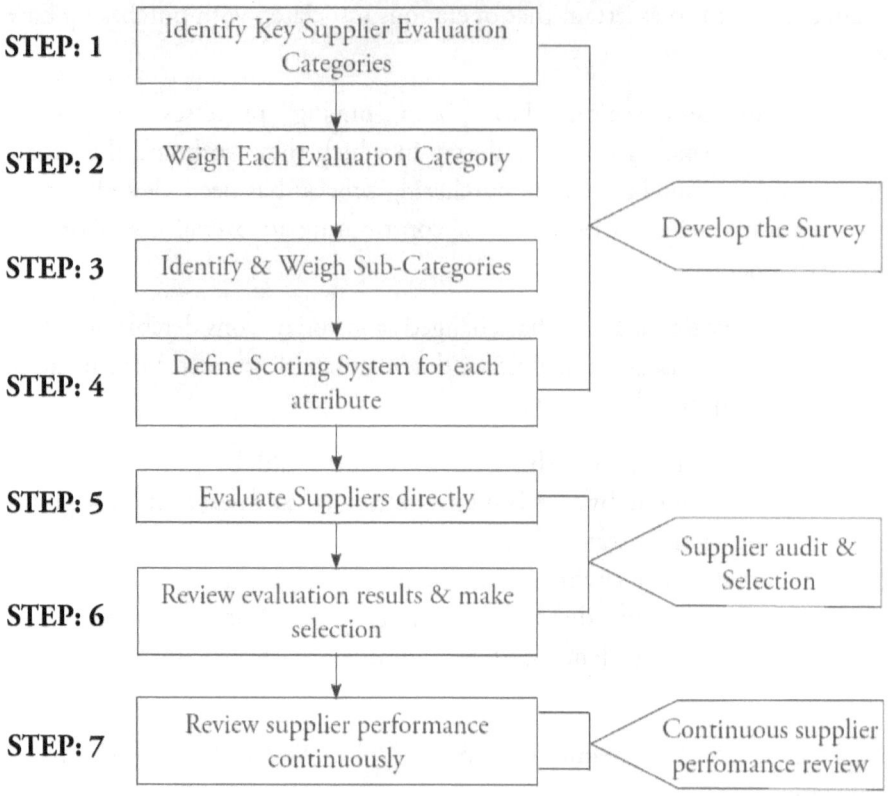

Step 1: One of the first steps when developing a supplier survey is for the purchaser to decide which performance categories to include. The primary criteria are cost/price, quality and delivery. These are generally the most obvious and most critical areas that affect buyers. However, for critical items needing an in-depth analysis of the supplier's capabilities, a more detailed supplier evaluation study is required which include Management Capability, Quality of People, Cost Structure, TQM and Process Capability, Regulatory Compliances, Financial Capability and Stability, Delivery Performance, Communication and IT Capability and Relationship Value.

Step 2: The performance categories usually receive a weight that reflects the relative importance of the category as well as importance to business in the supplier selection process. An important characteristic of an effective evaluation is its flexibility.

Step 3: This process requires identifying any performance sub-categories and distinguishing the same with different weight system if they exist within each broader performance category.

Step 4: A clearly defined scoring system takes criteria that may be highly subjective and develops a quantitative scale for measurement. Scoring metrics are effective if different individuals interpret and score the same performance categories under review.

Step 5: A buyer can select one supplier over another based on the evaluation score. It is also possible, based on the evaluation that a supplier does not qualify at this time for further purchase consideration. Purchasers should have minimum acceptable performance requirements that suppliers must satisfy before they can become part of the supply base.

Step 6: The purpose of the evaluation is to qualify potential suppliers for current or expected future purchase contracts. The primary output from this step is a recommendation about whether to accept a supplier for a business.

Step 7: When a purchaser decides to select a supplier, the supplier must then perform according to the purchaser's requirements. The emphasis shifts from the initial evaluation and selection of suppliers to continuous improvement by suppliers.

The use of weights and point should be simple enough so that each individual involved in the evaluation understands the mechanics of the scoring and selection process.

A survey conducted on 61 chemicals companies in India based on 39 criteria for evaluating suppliers and based on these criteria, an attempt was made to identify main criteria and sub-criteria. Moreover, these criteria are also prioritised based on management's perspective, the buyer's perspective and customer's perspective. The outcome indicates that quality-related issues dominated the decision-making process in the chemical industry.

Supplier selection process is also based on the complexity of buying commodity, For instance, if it is the selection of suppliers within "Strategic Segment," all aspects are investigated and due care is taken to ensure the sustainability of the relationship whereas, for other segments, a few criteria can be relaxed, or weight/importance of particular survey attribute can be changed accordingly. For a better understanding of the effect of "Market Risk versus Value Purchase" in the supplier selection process, the details of "Portfolio Analysis" are given.

Based on the spend quadrant (as per the portfolio analysis), the supplier selection cycle consists of eight steps:

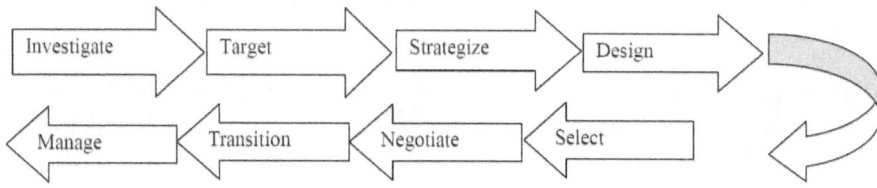

Revenue growth with margin improvement is the first priority for any business head. These tasks become challenging when customer demands changes, supply-chain is interrupted, the cost of material and the market competitive landscape becomes unfavourable.

The following countries are typically considered to be emerging markets which offer low-cost environment*:

1. China, Thailand, Vietnam

2. India, Middle East

3. Ukraine, Romania, Bulgaria

4. Mexico, Brazil

Implementing sourcing programmes in economically emerging regions is one way to grow revenue because, by sourcing goods in the local market, the company can compete more effectively and expand its business.

To reduce costs across the company, it decides to move 30% of the current expenditure on direct materials to low-cost country sources. Once this decision is taken, the company must determine which products should be sourced in low-cost regions and then determine the best suppliers. The following questions need to be answered:

- Who are the right suppliers?

- What is the SWOT analysis for product coverage in supplier's country?

- How can savings be ensured?

- How can a business case be built?

- How can product quality be maintained?

- How can the extended supply-chain be most effectively managed?

- How can a competitive lead time be ensured?

- How are the contracts set-up?

- What are the legal requirements?

- What are the benefits and pitfalls of local taxations?

- How can the local cultural challenges be managed?

- Does the supplier work for competition?

- Does the supplier demonstrate business ethics in line with expectations?

- Is supplier willing to provide reference customers?

Answers to the above questions will help to narrow down the list of potential suppliers. On the other side, careful portfolio analysis of product and market complexity also helps in taking the right and strategic decision from sourcing stand-point and many of questions can be answered through appropriate analysis of the purchase product and services versus difficulty in supply market. Portfolio analysis is a method of categorising a firm's products according to their relative competitive position and business growth rate in order to lay the foundation for sound strategic planning, where right supplier selection is the first objective and requires a systematic process with defined steps, i.e., "What must be done here to overcome the situation?", which is in the interest of business from a long-term perspective. This is also an integral part of strategic planning before the organisation goes for addition or deletion or change of suppliers with respect to the importance of purchasing versus market complexity.

When performing international sourcing and supply-chain management, "The Human Factor" has decisive importance. It comprises culture, trust, language, personal relationship and human resources. Part of "The Human Factor" is also the relationship between the sourcing company and the supplier.

Relationship plays a vital role when conducting business in India or so to say in any Asian country. Unlike the western world, Asians give more weight to "knowing the person" factor. In India, on the other hand, it is more important to meet the suppliers face-to-face where negotiations continue for a long period.

PROCESS FOR VENDOR IDENTIFICATION

Vendor identification starts with the search for potential suppliers. There are multiple ways to do this exercise, but the most common practices are as follows:

1. **Publication-Based Search**

    a.   Directories and Yellow pages.

    b.   Trade magazines published by professional bodies.

    c.   Geographic directories released by the Chamber of Commerce as well as other industrial associations.

2. **Specialised Sources**

    a.   Specialist press (International and Regional)

    b.   Trade shows

    c.   Consultants

3. **Informal Sources**

    a.   Buyers organisation

    b.   Personal and professional contacts

    c.   Competitors website

    d.   Supplier's website

During the course of vendor identification process, it is very important to ask information which may be required or relevant to our organisation in context to catering to short-term or long-term objectives if any association gets finalised, which requires setting up a structured questionnaire wherein, content is designed in such a way that buyer can assess whether the supplier has the minimum requirements to become a potential supplier.

For instance, if we want to develop a molecule wherein the involved reactions are chlorination, fluorination, distillation, etc. and it's very important to see the required reaction facility at commercial scale is available in-house for the sustainable competitiveness. It means along with general information no matter what kind of supplier, specific information/questions regarding specific commodity/segment, suitability to our requirement, i.e., technology, process, best practice, etc. are equally important.

The identification and selection of business associate or supplier are critical to the strategic sourcing process; sound partnerships are built upon reciprocity, commitment and shared vision by all engaged parties. The biggest challenge the sourcing leaders face is championing the art of identifying and selecting the right source for their organisation, which is not purely an art but science as well, as a lot of data collection, data analysis is involved in the entire process.

In order to help them and meet these challenges, it demands a new systematic approach to supplier development exercise.

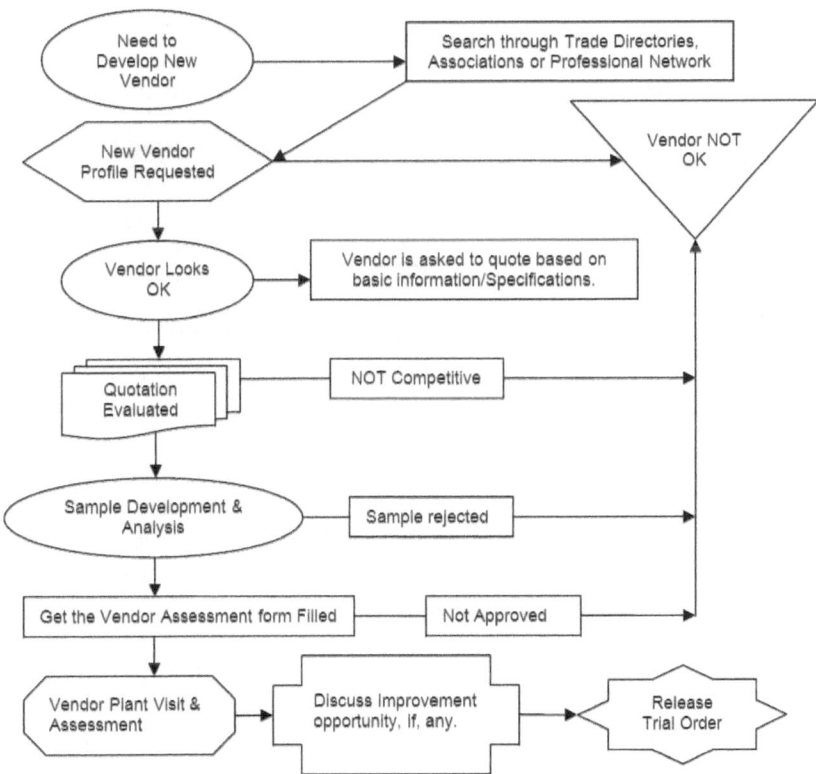

Most companies regard the use of supplier selection criteria as an important part of their supplier selection process. Supplier involvement in product development and continuous improvement efforts is much lower than the use of supplier selection criteria.

It is worth examining the number of suppliers needed. Buying from a carefully targeted group could have the following benefits:

- It will be easier to control suppliers.

- Business will become more important to them.

- They may be able to make deals that give an extra competitive advantage.

Further, when considering the supplier's organisation on our shortlist, ask the following questions:

- Can these suppliers deliver what we want, when we want it?

- Are they financially secure?

- How long have they been established?

- Do we know anyone who has used them and can recommend them?

- Are they on any approved supplier list from trade associations or government?

NEW SOURCE DEVELOPMENT PROCESS

While developing a vendor for a product, the major guidelines considered for selection are based on aspects related to the business model or organisation, either our own sourcing strategy or technical or commercial related guidelines.

In brief, while talking about organisation strategy-based guidelines, we focus on the financial status of the potential supplier, size of the company and its goodwill in the marketplace, ethical standard, IP compliance, company vision and value for this relationship.

Whereas, purchasing strategy-based guidelines focuses on supplier reliability, cost competitiveness, capacity and technical capability, geographic location, exposure to international marketing, supplier–supply-base model and logistics-related infrastructure.

Technical and process-based guidelines cover information on plant and machinery, process details, R&D capabilities, pilot plant related information, facility standard, patents and licences, quality and environmental related certifications as well as other engineering supports.

Whereas, it is equally important to screen the market for available suppliers based on certain criteria so that the selected suppliers have high possibilities of "staying-in" and meeting all expectations to become regular business associates or suppliers.

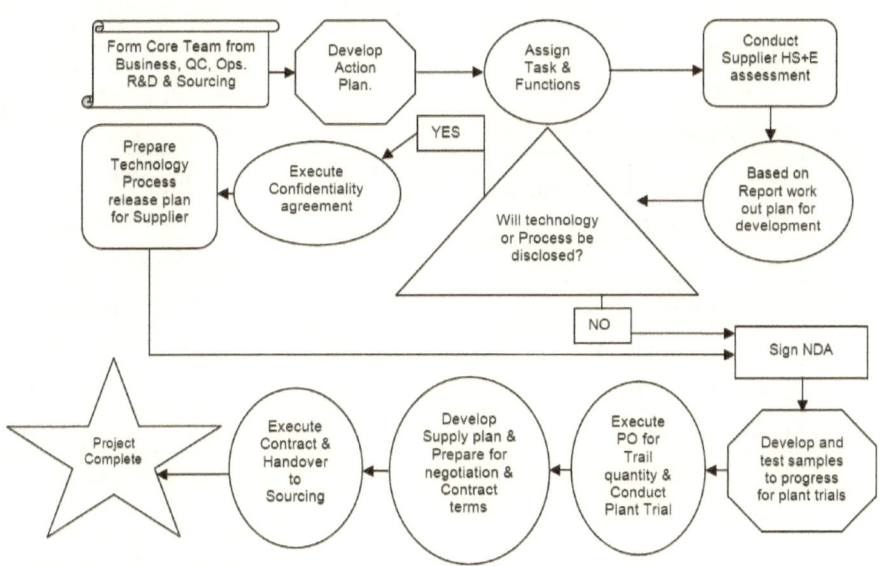

This trend has created a greater focus on selecting the right supplier. The traditional unit cost, lead time and quality have been replaced with the total cost, JIT delivery capability and use of TQM. A long-term relationship has driven new criteria such as financial stability, technical capability and organisational culture aspects. Trends in corporate relations have driven a new focus on the supplier's environmental standards and employee relations.

For instance, supplier selection methodology can be developed through creating the questionnaire based on participants' needs and requirements and then rating the importance of each attribute on a scale of 1 to 5 (where 1 means low and 5 means high). Weight is given to each attribute based on their criticality index and multiple of both these attributes are considered to determine the importance of the factor in supplier selection process.

KEY FACTORS TO PICK UP THE RIGHT SUPPLIER ORGANISATION

Organisational factors refer to the factors that might affect the different phases within the supplier selection process. In 1966, Dickson conducted a survey with 170 companies based out of the US and Canada from the much-diversified sector of the manufacturing industry. The result of the same was as follows,

Rank	Factors	Mean Rating	Evaluation
1	Quality	3.508	Extreme Importance
2	Delivery	3.417	Extreme Importance
3	Performance History	2.998	Extreme Importance
4	Warranties and Claim policies	2.849	Extreme Importance
5	Production facility and capacity	2.775	Considerable importance
6	Price	2.758	Considerable importance
7	Technical Capability	2.545	Considerable importance
8	Financial Position	2.514	Considerable importance
9	Procedural compliance	2.488	Considerable importance
10	Communication System	2.426	Considerable importance
11	Reputation and position in Industry	2.412	Considerable importance
12	Desire for business	2.256	Considerable importance
13	Management and Organisation	2.216	Considerable importance
14	Operating controls	2.211	Considerable importance
15	Repair services	2.187	Average Importance
16	Attitude	2.120	Average Importance
17	Impression	2.054	Average Importance
18	Packaging Ability	2.009	Average Importance
19	Labour relation record	2.003	Average Importance
20	Geographical location	1.872	Average Importance
21	Amount of past business	1.597	Average Importance
22	Training Aids	1.537	Average Importance
23	Reciprocal arrangements	0.610	Slight Importance

(Dickson's Results from the vendor Selection Criteria survey 1966)

Weber (1991), in his article "Vendor Selection criteria and methods", has made a significant contribution to understanding the earlier history of vendor selection criteria. Weber's principal critique of Dickson's work has been the lack of clarity in defining the criteria. For example, performance history may refer to delivery or quality, attitude and impression can be very subjective. Weber's work in 1991 has been based on the literature published on this subject up to 1991. Weber has taken Dickson's 23 criteria and determined their relative importance through counting the number of published articles referencing each of them. 74 articles were included in this review. Weber's results may be seen in the table below.

Rank	Factors	No. of Articles	Evaluation
1	Quality	40	Extreme Importance
2	Delivery	44	Considerable Importance
3	Performance History	7	Considerable Importance
4	Warranties and Claims	0	Considerable importance
5	Production Facility and capacity	23	Considerable importance
6	Net Price	61	Considerable importance
7	Technical Capability	15	Considerable importance
8	Financial Position	7	Considerable importance
9	Bidding procedural compliance	2	Average importance
10	Communication system	2	Average importance
11	Reputation and position in Industry	8	Average importance
12	Desire for business	1	Average importance
13	Management and organisation	10	Average importance
14	Operational control	3	Average importance
15	Repair Service	7	Average importance
16	Attitude	6	Average importance
17	Impression	2	Average importance
18	Packaging ability	3	Average importance
19	Labour relation record	2	Average importance

(Continued)

Rank	Factors	No. of Articles	Evaluation
20	Geographical location	16	Average importance
21	Amount of past business	1	Average importance
22	Training Aids	2	Average importance
23	Reciprocal arrangement	2	Slight importance

It is worth noting that there have been many changes in manufacturing in the 25 years (1966 to 1991) between these two pieces of research. The principal change has been in the introduction of JUST-IN-TIME (JIT) manufacturing where large inventories are to be avoided and shorter lead time is desired.

Weber states that given the complexity and economic importance of vendor selection and the multi-objective nature of this problem that multi-objective programming technique has not been put in use. He suggests that such techniques would allow purchasers to systemically examine the trade-offs among the criteria delivering the suppliers that best meet the company's needs.

Further, the supplier's organisation, depending on the strategic initiatives they have in place to build the supply-chain capabilities, have been segmented into three categories;

- **Operators:** This segment concentrates on cost reduction initiatives, process improvements and information linkages with key suppliers and logistics providers. They employ fixed structure/processes and focus on product flow with little integration. Here consensus plan is a single agreed-upon operational plan, which is developed by product development, sales and marketing, finance and supply-chain team members.

- **Planners:** Their strategies and initiatives characterise planning, operational efficiencies and cost reduction and containment. Enhance visibility to key partners with dashboards. Planners realise that the primary objective of the S&OP process is to create a single consensus plan translated into financial objectives that are executed across the entire organisation. They set up and position for growth through globalisation and customer alignment in the international market. They outsource their non-core functions to take advantage of logistics service providers and contract manufacturers assets and skills, technological and global footprint capabilities. With outsourcing, they are distributing or sharing the inherent risk of market volatility and demand variability by being in a better position to ramp up or down according to current market conditions.

- **Visionaries:** Their focused strategies and initiatives include supply-chain visibility with partner collaboration, business intelligence and analytics, risk management, optimisation of network, cost structure and inventory and customer demand management. They attack market volatility with analytical intelligence for supply/demand synchronisation and resource allocation. So, their major capability gets translated into the rapid response to changes in market condition and demand variability, use of market analytics and customer collaboration to predict demand, responsive allocation of all resources-human, assets, supply.

Based on the theoretical framework, the following external driving forces are identified in context to the chemical industry in India.

- **The Growth of Emerging Markets:** Well-developed infrastructure in India does not only make it easier to connect suppliers with customers but also to ensure the stability of supply channels and reliability of delivery. More and more qualified suppliers have emerged in India during the last decade due to the vast development of economy and technology especially pharma, agro and speciality intermediates in the chemical segment. Today, there are many manufacturers capable of providing standard products with high-quality.

- **Globalisation:** One reason why companies conduct sourcing in India is because their customers, suppliers and even competitors are already operating in India. Following the trend is an interesting aspect which argues that companies do not want to miss out on opportunities associated with global dynamic trends.

- **Liberalisation:** India is aware of her role in the global economy and has to loosen up earlier with tight regulations to promote international business as well as to meet western world business requirements to promote, export and earn foreign exchange. A more global, mature and liberalised Indian market becomes a solid base for competitive clusters.

Buying Factors

Buying factors refer to the factors that might affect the decision-making within an organisation and, furthermore, the supplier selection process. The factors that are considered in this section are the type of purchase and time pressure.

A recent Bain & Company survey of 138 manufacturing executives, in sectors ranging from automotive and chemicals to consumer products and technology, found that more than 80% of respondents believed that moving costs to low-cost countries was a high priority.

However, less than two-thirds had made it a significant company initiative and only 15% saw the benefits of offshoring value-added activities like R&D. Such incomplete efforts can change the benefits that firms seek by moving costs abroad.

To attack the problem from all sides, companies must go beyond whether to act and face the fact that cost migration is a competitive necessity. Forrester Research predicts that up to 3.4 million service jobs will move offshore. The Bain & Company research finds that such moves are netting manufacturers in Europe and North America cost saving of 20% to 60%.

The key to success lies in answering three other critical questions—**what, where** and **from whom to buy** or **how to migrate**.

For industrial companies today, shifting their cost base to low-cost countries is not an option but a competitive necessity. Hence, it becomes more important for local Indian companies to prepare themselves in a way that they can fulfil all relevant expectations of global companies to groom themselves as global suppliers.

A study conducted in this regard reflects the fact that 69% companies intend to lower cost by 10%, by developing and selecting supplier in low-cost countries, whereas 28% have a target of 20% of costs or more for migration to LCC suppliers.

More recent studies have discovered a shift away from price as a primary determinant of supplier selection. The set of relevant supplier selection criteria changes over time as a natural adaptation to the changing business climates and competitive environments.

For instance, it is more likely that buyers who are involved in a strategic partnership will not solely rely on traditional selection criteria and instead involve new selection criteria based on the nature of their requirement and priorities. This implies that probably there cannot be a generalised consensus on how to weigh the relative relevance of the different criteria.

How do sales and marketing people target their customers? Generally, sales and marketing managers have developed ways to divide their markets and allocate resources and to organise their sales forces and supporting logistics functions.

Moreover, while the search goes on at the front end of organisations to find even more subtle ways of understanding what the customer wants, it is important to reconnect the supply side to the demand side of enterprise supply-chains. Each is vital to the other as we seek a full and uninterrupted view of what is going on among our multiple chains. Because we are customers of the supply base in this situation, it's even more important to listen to our suppliers.

As emerging markets are typically more volatile and more competitive, mastering differentiating capabilities plays a significant role in success, as it supports rapidly implementing best practices without going through the painful learning curve many mature markets' companies have endured. Many emerging market leaders are leading the way by introducing innovative supply-chain practices to the global supply-chain community, especially in the area of sourcing flexibility and cost efficiency.

Value Driver	Top Three Differentiating Practices
Maximise delivery performance	1. Collaboration with key customers on planning. 2. End-to-end supply-chain planning and visibility. 3. Vendor managed inventory, direct replenishment model.
Minimised Costs	1. Best cost country sourcing. 2. Differentiated order to deliver on time. 3. Differentiated service level.
Maximum volume flexibility and responsiveness	1. Internal capacity flexibility 80% – 120% 2. Flexible shift models/payment structure. 3. Regional supply-chain set-up.

(Continued)

Value Driver	Top Three Differentiating Practices
Minimised Risk	1. Multiplication of sources and sole-source avoidance. 2. Regular review of supplier's financial risk and mitigation through risk sharing partnership. 3. Visibility and regular monitoring of main suppliers operational indicators.
Complexity Management	1. Development of multi-skilled employees to cope with complexity. 2. Late-stage product customisation. 3. Use of distributors and other channel partners.
Sustainability	1. Agreement with supply-chain partners to adhere to highest ethical standard. 2. Responsible supply-chain partner footprint and procurement framework. 3. Internal carbon footprint optimisation and improvement.
Tax optimisation and efficiency	1. Manufacturing optimisation–toll and contract manufacturing. 2. The localisation of inventory ownership in tax efficient countries. 3. The localisation of procurement organisation in tax efficient countries.

Organisations are paying closer attention to the concept of the sustainable supply-chain. Managing a supply-chain in a sustainable manner entails taking account of the impact of major environmental, social and economic factors throughout the life cycle of a product. In the long term, it's also what gives a company its licence to operate. Their survey has four main reasons for investing in sustainable supply-chain management—to manage the risk of unintended environmental or social damage, to manage organisation reputation and expectations of shareholders, to reduce cost and realise productivity improvements and to create a sustainable product, thereby increasing revenues and enhancing the brand value.

Based on the PwC 2013, a sustainable supply-chain requires collaboration with suppliers. As per the major practices for this segment, optimising internal carbon footprint is the top priority for 87%. Similarly, 87% favours adhering to the highest ethical standards, 81% favours collaborating with suppliers to create a responsible supply-chain footprint procurement framework, and for about 71%, effective track-and-trace capabilities are important.

As public expectations rise, companies will come under an increasing pressure to report on the environmental and social impacts of their activities and on the steps taken to mitigate those impacts.

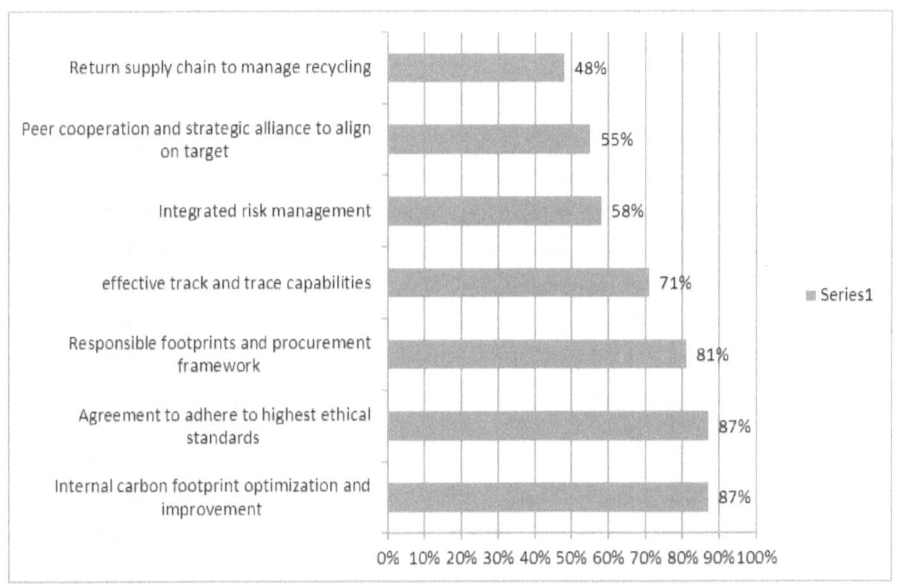

(Important practice for sustainability adherence, PwC global supply-chain survey 2013)

The following aspects enhance the value of buying factors while selecting the source.

- What are the supplier cost containment strategies? Is their cost structure fixed or variable?

- Is the supplier focusing on core capabilities? Are they outsourcing the right function? Are they taking advantage of the global cost, capabilities, local regulatory knowledge and skill of partner companies?

- Do they dynamically allocate all their resources: human, assets, supply and production?

- How is the risk factored into their operational decision-making and contingency planning? How do they measure the effectiveness of the risk management strategy?

- Do they have a sustainable strategy, reflected in product process and packaging, collaboration with customer initiatives and supplier compliance programmes?

Today's global marketplace is going to become even more competitive over the next few years, as organisations seek to optimise their supply-chain and procurement to respond to constant demand variance, adopting new rules with respect to supplier selection to restore stability to supply-chain/procurement function.

As we know, there are several buying factors such as lower prices, quality, technology access, access to new markets, shorter product development and life cycles, comparative advantages. However, adoption of global sourcing often leads to some problems such as increase in inventory level because the average inventory level depends on the supply lead time and variability, which get higher when the distance between supplier and buyer increases, there comes the need for Vendor Managed Inventory (VMI) which works as one of the major buying factors for new supplier or plays an important role in supplier selection factor.

VMI is an operating model in which the supplier delivers its goods to the customer, but the actual sale is postponed after the actual use/sale by the customer. The supplier which can be manufacturer, reseller or a distributor, monitors the customer's inventory levels and makes inventory replenishment decisions regarding order quantities, shipping and timings. VMI can be a major buying factor as it results in reduced inventory costs, better response to market changes, reduction in demand uncertainty and more flexibility in production planning and distribution.

Since globalisation has boosted markets competitiveness, industrial companies have relied on offshoring and outsourcing initiatives. Such initiatives have led to high network complexity and higher inter-dependency among supply-chain organisations, thus fostering supply-chain vulnerability and here VMI concept is seen as USP in the supplier selection process.

In short, buying factors' selection highlights the importance of choosing carefully supply market attributes keeping in mind local/global sourcing strategy and how it can support leveraging right supply-chain management tool.

GO GLOBAL AND ACT LOCAL

Global sourcing is a critical element for many multi or trans-national organisations in their value-added chain to be competitive at a global scale. It's primarily a threat from low-cost countries and emerging global cost competitiveness forcing large global companies to achieve added value in their transactions through either manufacturing, marketing or sourcing collaborations. Now, the overall trend is pushing even small or medium-size foreign companies towards low-cost international sources of supply.

Global sourcing has changed the face of sourcing/outsourcing in the last 7 to 10 years. This has created big industrial opportunities for countries like India, China, Brazil and Eastern European Countries. Today, most of the companies face the maximum intensity of competition from China and other low-cost countries.

For any company outsourcing means replacing the in-house production of an intermediate or final product with that of purchased or sourced from outside the company/country and is an integral part of value-added chain. While the word "global sourcing" is looked at as a cost improvement exercise by many, the other aspects which need attention are the substantial amount of risk and complexities. Global sourcing adds to the value chain. The list of risks and complexities are as follows:

- Long cycle time
- Ambiguous total landed cost
- High inventory risk
- High regulatory risk
- Rejection related disposal cost
- High intellectual proprietary risk
- High overhead cost
- Information availability is low
- Reliability related risk
- Political uncertainty related issues

- Culture and communication-related gaps

- Processes and security-related risk is also from moderate to high

Therefore, for any company willing to outsource part of its value-added chain, it's important to ensure that their supplier selection process in the global sourcing function addresses all these risks and adequate provisions or processes are in place to safeguard the buyer's long-term business interests. For instance, detail working on overall storage and logistics-related cost involved with the supply-chain cost while comparing the sourcing cost from local source versus overseas supplier, product price benefit versus increase in inventory cost and increase in rejection risk, increase in lead time involved in the procurement process as well as substantial increase in overhead cost associated with additional infrastructure, people and other indirect cost involved with running these global sourcing offices outside the country.

Therefore, it becomes equally important for a foreign buyer to evolve a process of source selection in context to global sourcing which has provision to deal with intense global competition, sustainability of the relationship, thrust for continuous cost improvement, total quality excellence in order to make this new association a win-win equation for seller as well as for buyer.

The case study shared in this book is in context to Specialty Chemicals manufacturing companies, where all they should focus in order to be competitive is to bring big sourcing/outsourcing opportunities to their business portfolio and join hands with global chemical producing companies. These attributes are evolved based on the assessment carried out in numbers of speciality chemicals manufacturing companies from emerging regions.

There are different approaches possible when deciding how the research should be conducted. The first step is to determine if the research should have a deductive approach or an inductive approach and then decide whether the approach should be of quantitative or qualitative character. The deductive approach means that you first develop a hypothesis or a problem formulation and then design a research strategy to test it in real life. Whereas, in the inductive approach you collect data and then develop a theory based on the data collected. This study follows an inductive approach based on quantitative data in order to establish correlations with the use of statistics.

To provide some context the following is a summary of the process used with selected chemical manufacturing companies in India. The size of the companies ranging from $100,000 sales turnover to over a billion USD sales turnover. Companies were also screened on the following parameters;

a. Number of suppliers added per year.

b. Number of suppliers discontinued per year.

c. Number of new products/molecules developed per year.

d. Number of detailed supplier assessment conducted per year.

To facilitate an easy comparison between companies, this survey was divided into 3 parts. Part I of the survey primarily focuses on Management Perspective, i.e., the importance of various criteria on a scale of 1 to 10 for seven attributes, which the management of supplier's organisation thinks are important to them in their business alliance or supply partner.

Part II of the survey focuses purely on procurement objectives, wherein this survey tries to identify important criteria for source selection by seeking weight on a scale of 1 to 5 From buying team of supplier company. Each respondent-buyer was asked to rate key supplier selection criteria on a five-point scale from 1 (lowest) to 5 (highest).

Part III is more from Sales Perspective, which means seeking information from the S&M team of supplier's organisation on important criteria where their customers take the given attribute in the decision-making process.

The following attributes were considered for assessment and discussion with companies which affect their supplier selection process. To identify and prioritise the criteria, interviews were done with the buyers in chemical companies of various sizes and segments.

In line with these interviews, seven main criteria were identified. Further, thirty-nine sub-criteria were worked out within above main criteria which affect the supplier selection process largely within chemical companies from buyer's perspective. While making the selection, it is important which criteria are to be selected for the best solution. The perspective and importance of these attributes are explained below:

- **Goodwill and Reputation**

  - *"Organisation with a vision to grow, valuing people and community."* The sourcing (or buyer) company, while assessing this attribute focuses on whether the leadership of potential supplier has the same vision as them and whether the potential supplier is willing to put their business as the first priority for them in terms of resource deployment and usage. Management stability of the supplier is also considered to be an essential criterion in vendor selection. It is because the long-term relationship requires consistent commitment from the supplier's management.

    Also, the buyer looks at whether the supplier sees value in building a long-term business relationship, as the buyer will also be taking a business-related risk in terms of time and money by grooming new source in order to come up to level to match their requirement, which normally is a complex equation. At the same time, the buyer also tries to assess whether the potential supplier or manufacturer has a good market reputation with his existing vendor or supplier base. Whether there are system and top management commitment in place for compliance of law and other business ethical standards in context to the protection of proprietary information.

    - *Proprietary Information and Protection*

      - *"Has been working with several multi/trans-national companies with significant technology shared by buyers and has a successful track record on technology JV's."*

        No local version of product introduced. Good practice in place to maintain access to proprietary information. The supplier has internal robust systems like coding, etc. to protect the buyer's IP in addition to a signed confidentiality agreement with each key employee. It is applicable even after the employees quit the organisation.

■ *Company Vision Alignment*

☐ When Alliance Partner or Buyer–Seller organisation corporate values are congruent with each other's company values, their relationships are better and lead to higher confidence and better productivity. Does the supplier communicate its vision of the potential reward for both parties and have a coherent process for achieving it?

■ *Not a Generic Competitor*

☐ Envisages as an intermediate producer in the present and near future, significant revenue is from tolling business. Establishes well in current business and would like to grow in same, multiple operation units serving varied industry and believes in serving all.

■ *Historical Relationship*

☐ *"Has a working relationship with the buyer, has been supplying to the buyer through other agencies in the past."* Supplier or Supplier Group Company is known to the customer in the past. Even knowing an individual or knowing leadership in the past helps in building confidence w.r.t. newly formed relationship.

■ *Value Is Seen in Association/Desire for Business*

☐ *"Share same values on business ethics and practice, working with buyer adds credibility in the marketplace and current supplier."* This is also how customer-focused the supplier is. What kind of flexibility as a supplier are they willing to accommodate to win this new business relationship? This affects the supplier ability in context to service level, functionality and costs. The critical driving force is the future of business opportunities where companies benefit from extending relations with important partners. During the past few decades, many European companies have moved their manufacturing function partially or entirely to India.

■ **Communication System**

☐ The components of a communication system serve a common purpose, are technically compatible, use common procedures, respond to controls and operate in unison. A strong communication system supports follow-up, internal coordination, accessibility and ease to communicate which helps in faster and timely decision-making.

● **People**

○ *Quality of people is the greatest and most important asset for any organisation;* this attribute also affects the growth and sustainability of any new association from a buyer perspective. It becomes necessary for the buyer to know the background of the people from the supplier company if the association they are targeting to form is of a long-term nature. At the same time, it is also important to assess the external resources, which are in place to carry out other business-related activities. The organisation's willingness to change for continuous improvement in their existing process of management as well as whether there is commitment and system in place to take care of issues related with compliance of Health, Safety, Environment and Responsible Care related activities. Quality of people also takes into account academic profile, relevant experience and presence of teamwork to achieve a common target for the organisation.

■ **Labour Relations Record**

☐ An employer's level of power over its workers is dependent upon numerous factors, the most influential being the nature of the contractual relationship between the two. This relationship is affected by three significant factors: Interest, Control and Motivation. It is generally considered the employer's responsibility to manage and balance these factors in a way that enables a harmonious and productive working relationship. Employer and managerial control within an organisation rest at many levels and have important implications for staff and productivity alike, with control forming the fundamental link between desired outcomes and actual processes. Employers must balance interests such as decreasing wage constraints with a maximisation of labour productivity in order to achieve a profitable and productive employment relationship. There is no record on labour unrest or strike, etc. in the past and how frequently they meet to resolve issues if any.

■ *Training and Orientation*

□ New employee orientation effectively integrates the new employee into the organisation and assists with retention, motivation, job satisfaction and quickly enabling everyone to become contributing members of the work team. If done correctly, new employee orientation can solidify the new employee's relationship with the organisation.

■ *Willingness to Change/Flexibility of the Organisation*

□ **"This company is willing to work together for improvements/change. This company historically has met deadlines in spite of changes in protocol during the project."** Flexible on Business Terms/ Volumes/Process Changes/Upgrades/Schedules, etc. Customers must consider the degree of flexibility they require from the supplier. Flexibility is an important aspect, especially when companies perform offshore outsourcing. One of the main motives for enhanced flexibility is the ability to change according to market demands. An opportunity where companies could free up their own resources from outsourcing functions to others and focus on what they do best to ensure a wide range of product portfolio.

■ *Organisation Efficiency and Response*

□ **"This company uses the matrix management concept with decision-making delegated to appropriate Supervisory Levels."** This is the capability to deliver "Win/Win" results for customer and supplier, which cannot be possible without properly skilled, experienced and motivated people, who are fundamental components for organisation efficiency. Organisational efficiency is based on the supplier's ability to deliver radically improved services in terms of cost and quality. It mainly involves the supplier's leadership, sourcing, process improvement, technology and customer development capabilities. Supplier's adequate resources are equally important to improve their efficiency like Energy supply, transportation, the physical building, telecom, IT infrastructure. Professionalism including sub-criteria like accuracy, expertise, attitude and reliability is critical to the success and mutually beneficial business relationship.

■ **Quality of Human Resources**

□ "This Company has adequate human assets for its current practising Chemistry/Process. This Company has had X% of Human Asset Turnover and has managed well to maintain its standards/competencies." The wise organisation will have the right mix of capabilities because personnel from the supplier organisation are those who guarantee the business.

■ **Leadership Commitment**

□ Company has well-defined and capable leadership to promote and progress with this project. Company's leadership believes in the value this project has and is ready to allocate sufficient resources to meet the deadlines. Company's leadership is compatible/ aligned with the buyer on Project Prioritisation. Leadership is the capability to identify and deliver overall success throughout the deal. Leadership commitment helps in building a strong relationship with the customer side organisation. This helps in delivering in line with the agreements, the supplier's and the customer's business plans.

- **Service Level/Delivery and Logistics**

  ○ *Catering to the customer on time is the key to success.*
  Right infrastructure support to ensure timely development
  and delivery of samples, as well as adherence to committed
  schedules, go a long way in strengthening the relationship.
  Chemical companies need to be equally good in case of the
  sample quality as well as the quality consistency at a commercial
  scale of manufacturing, which was often considered a gap.
  It is the cost and competition which is forcing businesses to
  look at low-cost countries for the supply of their raw materials
  and intermediates to retain their competitiveness. Therefore,
  it becomes very important for the buyer to ensure during
  the selection of a new source that potential supplier offer the
  most competitive price and product quality. Whether there
  is a system in place in context to documentation compliance,
  international logistics-related in-house competencies,
  commitment towards lead time, the location of the plant and
  international trading experience related with hazardous nature
  of chemicals are also equally important. Delivery is another
  critical factor in the supplier selection model. Geographical
  location, political environment, as well as the economic
  environment, also play an important role in the selection
  process. For instance, during the Beijing Olympics in 2008, to
  reduce pollution and traffic congestion before, during and after
  the games, the Chinese government implemented a variety of
  measures to restrict and close down heavy pollution industries.
  These measures had a detrimental effect on international trade,
  as the industry was not able to fulfil its supply and delivery
  related commitment to domestic or international customers
  for a good amount of few months, which shook reliability
  aspects for international customers, in perspective to related
  "geopolitical" reason which is beyond their control but affects
  the company's service performance level.

  - *Reliability to Deliver and Perform*

    □ In terms of the compound supplied to the chemical
    industry, it is important that the compound is delivered

to the customer with no structural deterioration or partial damage, that is, with no compromise from efficacy. Appropriate transportation and speedy delivery play an important role in selecting a supplier. "In two out of four cases, delivery deadlines were not met within the last 2 years. Has a process in place to measure and improve Customer Order Delivery Performance." The delivery competency is based on the supplier's ability and willingness to respond to customer's day-to-day operational needs. This studies consistency in meeting delivery deadlines, order fill rate, flexibility in meeting client's special delivery requirements.

- ### Lead Time

  □ "Sufficient capacity/de-bottleneck process to deliver on time." In production planning, lead time is very important. Suppose, A product is made from material B and material C. Material B is made from material C. It means, if there is a delay in the supply of material C, there will be the delay in production of material B. Eventually, there will be a delay in the production of product A. Customer will not get delivery on time. The demand-driven supply-chain requires a flexible production plan that requires materials to meet this plan in the shortest time possible.

- ### Plant Location

  □ Plant location plays an important role in overall transportation cost. Transportation cost is a criterion that both firms and suppliers are very sensitive to. Transportation cost is not merely the cost of sending the goods from one place to another, but also covers the transportation of the goods under appropriate conditions and in an appropriate way. Careful analysis is required in this case to eliminate the risk of disruption in the supply-chain. In international sourcing perceived international risks mainly consist of geopolitical risks.

- **■ *Warehousing/On Time to Request***

  - ☐ As manufacturers reduce their inventories of raw materials, the concept of on-time delivery has become more and more critical. Often manufacturing plants hold only a few days' material stocks. Without the buffer of a large stock of raw materials, a delay in delivery of materials can result in lines shutting down due to lack of materials and missing commitments to the manufacturer's customers. Reliable deliveries are critical. The cost of the high inventory and risk obsolescence must be considered when agreeing on a minimum order quantity with the supplier.

- **■ *Product Availability***

  - ☐ "Two out of four cases delivery deadlines were not met within last 2 years. Is the process in place to measure and improve Customer Order Delivery Performance?" The demand-driven supply-chain requires a flexible production plan that requires materials to meet this plan in the shortest time possible. The time from placement of materials order to delivery if, higher will become a constraint on the manufacturer's production schedule and hinder responsiveness to the customer's demand. This is also the capability to track and measure supplier performance.

- **■ *Procedural Compliance***

  - ☐ The chemical industry is under tremendous pressure for the developed world towards compliance of rules, regulations and documentations w.r.t. country-specific requirements for the involved molecule for instance REACh (Registration, Evaluation, and Authorisation of Chemicals) regulations for Europe. This also means that the manufacturer or interested suppliers who are in compliance with REACh rules and regulations can only supply to Europe.

- **Quality Systems**

  - *A quality management system (QMS) can be expressed as the organisational structure, procedures, processes and resources needed to implement quality management.* Quality is an integral concept that is related to all the steps in the value chain. Performance of quality improvement activities follows up of quality costs and availability of a quality control database are the main indicators of the supplier's success in quality planning. Supplier Quality Management is a core issue since final product quality affects not only the relationship between the supplier and the buyer but the whole supply chain. The vendor organisation corporate culture, i.e., commitment for providing high-quality products and for preventing quality failures also has an impact on this criterion. This should include quality certificates, control procedures and quality assurance, complaint handling procedures, quality manuals and internal rating and reporting systems. Sourcing organisation may tend to have suppliers using either the international standards or the standards sourcing organisation are currently using. Whether or not the potential supplier has a good track record in doing international business is also considered to be an important criterion relevant to internationalisation.

    - **Product Quality and Consistency**

      - The main reason the company prefers a global supplier is that this enables them to reach high-quality products. Within the quality criteria, purity profile is perhaps as important as price for the chemical industry, as the purity of the compound increases the efficacy of the produced also increases. Hence, suppliers providing high and consistent quality are likely to be preferred. Compliance with GMP standards and ISO documentation, i.e., clean-out procedures and practice is in place. In general, the company recognises 2 aspects of the cost of quality. First, the cost of potentially lost businesses due to poor quality and second, the cost of rework/repair, etc.

- **R&D and Analytical Capability**

  □ "R&D facility is equipped to absorb as well as improve upon the present technology. For example, the Analytical facility has all the latest equipment for conducting a test." Supplier R&D and Analytical capability is also detrimental in context to testing costs as well as development and process improvement perspective. Good R&D and the analytical facility can be helpful w.r.t. continuous process improvement and staying competitive in a global environment.

- **Product Performance**

  □ Product performance refers to the process where the supplier guarantees product reliability, produces the product with requested specifications and within the time frame specified by the customer, delivers the product in good condition and in the specified time with no loss in technical attributes. This aspect is also very well connected with domain expertise; it is the capability to retain and apply knowledge. The key here is not just a supplier's technical know-how but the ability to acquire and understand the business experience in a customer's specific kind of sectoral environment. If the purchased product is designed for the end user, compatibility, durability, functionality, maintainability, reliability and quality needed to be included in the product performance criteria.

- **Warranties and Claim Policy**

  □ It is also important that the compound carries the conditions and properties demanded by the customer at the time of delivery. A supplier's ability to offer warranty and insurance terms to protect the product against any problem that may be encountered at this stage is highly effective in choosing a supplier. This also creates a higher confidence level for offshore and domestic customers in the supplier of their product and their commitment to meet customer's quality requirements.

- ***Packaging Ability***

  □ Correct packaging of the compound to be used in the chemical industry is highly important in preventing any deterioration in the chemical properties of the compound. While some chemicals require transportation and handling at certain temperatures some must never contact air or specific DG/HAZ category product requires specific packaging materials for AIR/SEA/ROAD transportation to avoid any environmental or human injury or accident or control spillage, etc. Hence, the packaging method selected by the supplier is very important and varies for each compound to ensure that the product reaches to the customer in safe and in right quality.

- ***ISO's and Other Quality Certifications***

  □ Product quality can be defined more clearly based on the organisation quality systems. It is important to choose suppliers with national and/or international quality certificates confirming their quality standards. ISO 14001 specifies the requirements of an environmental management system for small to large organisations. Conformity to MSDS (Materials Safety and DataSheet), which requires controls for handling and storing of especially dangerous and poisonous materials as well as waste treatment disposals are critical issues for management and selection of the right supplier.

- **Health, Safety, Environment and Social Responsibility**

  ○ ***Committed to meeting the highest standards of corporate citizenship by protecting the health and safety of employees, safeguarding the environment and by creating a long-lasting, positive impact on the communities where we do business.*** In chemical companies, commitment to the health of their employees and people staying in local vicinity, safety of their employees while they are on the job, ensure that organisation is not violating any environmental norms or not polluting the environment rather fulfilling as corporate citizen their social responsibility towards society and country as well as mother-earth and humanity at large. Environmental, economic and social responsibility is not simply referred to other agencies, but to see if the organisation is involved in people and community aspects. Since the environmental pressure is continuously increasing, sourcing company should integrate environmental criteria into the vendor selection process. Quantitative environmental criteria include the environmental costs which can further be decomposed into pollutant effects and improvement. Qualitative environmental criteria consist of five elements: management competencies, green image, design for environment (recycle, reuse, disposal, etc.), Environmental management system (ISO 14001 certification) and environmental competencies.

- ### *Compliance of Law and Ethical Standards*

    □ "Meet all requirements of law and business practice," has external audits on financial and other aspects of the business management process. Well reputed in local industry for maintaining business standards. There is no complaint or violation record with government and local bodies in terms of breaking the law and meeting all required business ethical standards for running their business on the day-to-day basis. This is one of the most important criteria for global customers as any violation can jeopardise the entire development cost.

- ### *Health, Safety and Environmental Compliance*

    □ Supplier's safety and environmental concerns are incorporated in the model due to the increasing legal as well as public pressure on the protection of environment and pollution. This company monitors and motivates its employees towards Safe Work Culture. This company can meet basic safety, Health and Environmental related Audit Protocol has Safety Performance Data. They work with the local community to maintain relations. Can handle safely toxic/highly toxic materials. Linking supplier selection to HS+E performance.

- ### *Signatory of Responsible Care Programme*

  - "It has adequate Waste Handling Processes/Equipment to meet current needs. It meets all relevant Local Regulatory Norms on all forms of Emissions and Permits." As a signatory of responsible care programme, supplier certifies their commitment to safety, health and environment in the chemical industry. While the dumping of waste into landfill was simply part of doing business in the 20th century one of the hallmarks of success in the 21st century will likely be a company's ability to avoid waste altogether.

- ### *Documentation Compliance and Knowledge*

  - "Has prior experience of Exports and Documentation needed for timely supplies. Currently, exports have a significant percentage of production." This will be helpful to ensure that the compliance aspect will be taken care of in a business transaction. At times, documentation requirement varies from country to country; hence, relevant documentation compliance knowledge will be the key aspect of a business.

- ### *Exposure to Related Regulations*

  - Customs, Tax and International trade laws that vary from country to country may lead to big problems for both the suppliers and the customers when the compound sent by the supplier without appropriate documents or under inappropriate conditions. The supplier must take these possibilities into account and have relevant measures in place to counter them. It is undeniable that government policies regarding exports also have an influence on the final performance of the outsourcing activities.

- **Product Registration and Handling**

  - ☐ Infrastructure, resource and experience to handle registration related issues with government sectors. In the chemical industry, be it Agro-Chemicals, Pharma or Food Additive, Product Registration and handling the same to ensure compliance is an important aspect and potential supplier's knowledge and experience in this is quite important.

- **Financials**

  - ○ *Financial viability and business diversity of the potential supplier is also one of the key factors that form strategic and long-term buyer–seller relationship.* Financial flexibility works as a support mechanism to acquire all those skills required to become business partners. While easy access to strategic resources such as talent, raw materials, etc. would be of great interest to any foreign company, if it is available with their low-cost region partner, for which adequate financial strength is the key. Financial stability could be a good indicator of the supplier's consistent performance. In other words, suppliers with a solid financial position are more likely to maintain the quality and service level required by the sourcing company in the long term. In international sourcing, currency exchange rate fluctuation is the major component of financial risks as it can increase the cost dramatically. The production cost with suppliers must be lower than their customers.

    - **Financial Soundness to Make Investment**

      - ☐ "Established company with high financial ratings." "Ready to share information about the financial soundness for the past three years." "Has managed borrowings well and got a good credit rating in the local environment." It is essential that vendors provide a consistent supply of materials. If the vendor is not financially stable, cash flow (and material flow) and possible closure will affect the material supply. The net effect of this scenario is a loss of investment costs when transferring business to this supplier and an interrupted supply of materials. This will also lead to the loss of future sales due to disappointed customers. The current global credit crunch has resulted in many companies going out of business due to credit and cash flow issues. For these reasons, it is essential that some measures are put in place to ensure that a new vendor can pay the bills and remain in business for medium to long term. Many organisations use the acid test or liquidity ratio (Current assets-stock/Current liabilities) to determine the liquidity of a company.

■ *Global Cost Competitiveness and Sustainability*

☐ For companies, it is very important that the total cost is kept low. Thus, sourcing organisations should thoroughly investigate the supplier's negotiability flexibility when evaluating potential suppliers. Otherwise, the long-term relationship will be undermined if the vendor remains stubborn during the negotiation process. The price of the product or a compound makes a significant part of the total cost. Hence, the low product price is an important reason for choosing a specific supplier. Further, in global competition, taxes vary both for the buyer and the seller. Hence the laws of the country where the selected supplier operates may bring an additional obligation which might get added onto the product cost. The final delivered cost is X% over/under current/target price. Falls in ballpark cost-based on Cost Modelling Tools like Reverse Auction.

The cost has traditionally been considered one of the most important aspects of supplier selection criteria in the purchasing and supply management literature. Other factors such as minimum order quantity, lead time and forecast volumes will influence the material unit price. Even where materials are quoted at an ex-works rate, the vendor must be assessed on the cost of materials delivered to the plant. John Ruskin (1819–1900) has made very valuable comments w.r.t. pricing which says, *"It is unwise to pay too much, but it is unwise to pay too little. When you pay too much, you lose a little money, that is all. When you pay too little, you sometimes lose everything, because the thing you bought was incapable of doing the thing you bought it to do…"*

- **INCO/Pay Terms Flexibility**

  □ Financial terms are critical criterions which directly affect the cost and profit levels. This affects the overall cost of the receiving products. The supplier should have adequate funds to complete the supplies and should be financially stable as it adds to supplier flexibility. Flexibility would facilitate the adaptation to changing demands and needs. The credit terms offered by vendors can have a direct effect on cash flow. Longer credit terms allow the opportunity to convert materials into finished products and then into cash before paying for these materials. So the terms of payment should also be one of the evaluation criteria in supplier selection.

- **Manufacturing and Innovations**

  o *This is more related with potential supplier's overall technical/ operation capability assessment,* in order to ensure that whether the manufacturer will be able to produce consistent quality of products matching specifications with all the necessary facilities in place along with required research and development support and technology related back up. Competencies related with registration, testing, etc. also play an important role in qualifying the vendor as a strategic partner. While performing the assessment of vendor, one of the tried and tested tools on this aspect for the buyer is to visit the quality control laboratory and audit each equipment for calibration schedule, documentation and record keeping in context to data on various batches analysis and testing, etc. before spending the time on plant visit. The quality lab is the cockpit of the plant. If that is not in order, there are high possibilities that the plant and other things would not be in an organised manner. If innovation were considered a simple process with input and outputs, it may have been measured with less difficulty. The key input into innovation must be resources in training and R&D. The typical outputs of innovation are new products and intellectual property.

■ *Manufacturing Capacity and Capability*

  □ It is important that the supplier is not only able to supply the current demands of the customers but also able to readily adapt and quickly deliver the future demand that may come from the customers. "Size of pots and pans, existing infrastructure is capable of handling volumes in question without major investments." This also defines supplier capability to deliver inter-related projects. Strong supplier capability in manufacturing management can influence a customer's decision to expand the use of the supplier.

■ *Contamination and Prevention*

  □ "CP documentation. i.e., clean-out procedures and practice is in place." The effectiveness of the quality management system can be best measured by the number of complaints, and the same can be effectively managed with the effective programme on contamination and prevention. The company's SOP (Standard Operating Procedure) on contamination and prevention and its execution on a routine basis is a significant trait the customers will look for in their suppliers in the chemical industry.

■ *Continuous Improvement Programme*

  □ The focus should be on exceeding customer's expectations. Suppliers, particularly those operating in the chemical industry, should watch their customers well and be able to identify and produce in good time the compounds that the customer may demand. Continuous improvement programmes like Kaizen and Six Sigma are tools for improving standardised activities and productivity. They are vital considering today's fast-changing customer requirements along with high level of competition. This is a value system of the organisation which defines capability to incorporate changes to the process to meet dynamic improvement targets.

■ ***Process and Facility Standard***

    □ *"GMP* practice and technological standard of the plant." Process capability is significant since it assists in investigating the supplier's capability to produce quality products. The output of an in-control process is compared to the specification limits using capability indices and the process's ability to create a product within specification limits is measured effectively. Therefore, process and facility standards are important aspects to surface qualitative and competitive suppliers.

■ ***Technical Expertise/NPD***

    □ Engineering and technological standard of Plant/ Organisation. This is the capability to swiftly and effectively deploy new technology to support new product development at a desired level of speed. This is one of the major reasons why customer outsources product process, i.e., N–1 or N+1 level of chemistry to harness supplier's capability and investment. Investments that customers were unable or unwilling to make.

■ ***Technology and Engineering Support***

    □ Supplier track records can provide vital information on re-engineering for customer's as well as the supplier's skills and change capability to meet customer need and requirements with respect to plant engineering and technology. The sourcing organisation should examine suppliers' company technical competency based on accessibility, timeliness, responsiveness, dependability in context to if they need technical support in order to consistently provide high-quality product/service and to promote the developments and improvements. Moreover, technical criteria weight more when the client organisation is on track for developing a new product.

The concept of reliability is related to the rigour with which the task of data collection and analysis and the care with which report describes in detail each parameter meaning and intent as done for the purpose of understanding and supplier evaluation/screening, including discussions with respondents to complement survey data and analysis. Here, often the sensing reliability is equated with the term methodological accuracy. All the results and expressions during the discussion are analysed manually is shared in the subsequent chapters, which to a great extent depends on the surveyor understanding of the customer focus, market dynamics, communication effectiveness and understanding of the respondent of the importance of above-stated criteria in their supplier selection process.

REPRESENT INDUSTRY MIX IN SUPPLIER SELECTION PROCESS

Compared to MNC's, the role of SMEs in global economies has drastically increased in the last decade. In general, there are two main types of sample selections to use: probability or non-probability sampling method. A probability sample is a selection of sampling techniques in which the probability of each case being selected is known and is often equal for all cases. A non-probability sample is, on the contrary, a selection of techniques where the probability of each case being selected is unknown. Occasionally, it may be possible to collect data from every possible case in the population, and this is termed census. The sample in this study can be considered a census, considering that representative samples of small, medium and large size organisations from Indian chemical industry are included in this survey. This is based on 61 organisations that were included in this study.

Much of the data gathered in this report are based on discussion with respective participants, e.g., the size of the organisation, location, year of establishment, sales turnover as well as % of export to overseas markets. 100% of the respondents selected are chemical manufacturers either for the speciality chemical industry, Pharma, Agro-Chemicals or Polymers industries. Sixty-one companies were selected for this survey representing small to medium to large-scale companies in the chemical industry and senior executives as well as procurement professionals of these participant companies were interviewed.

- Profile of Respondents
    - i.   % Small-scale companies (Up to $ 2 Mn) – 18%
    - ii. % Small to Medium scale companies ($ 2 Mn to $ 10 Mn) – 23%
    - iii. % Medium scale companies ($ 10 Mn to $ 50 Mn) – 34%
    - iv. % Medium to Large-scale companies ($ 50 Mn–$ 100 Mn) – 8%
    - v.   % Large-scale companies (Above $ 100 Mn) – 17%

The largest part of study respondents associated with offshore outsourcing/sourcing activities to cater to their MNC's customers, who have long experiences of moving activities abroad to seek additional advantages.

Consequently, studying SMEs as well as large companies becomes more interesting due to lesser research on why these companies conduct international sourcing including vendor selection.

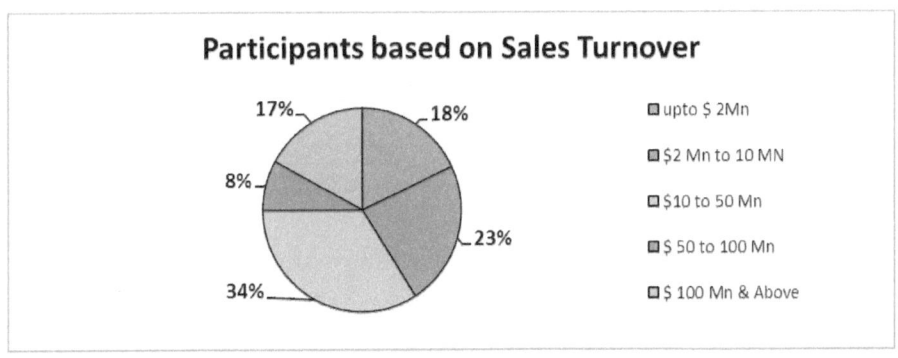

Participants based on Sales Turnover

- upto $ 2Mn
- $2 Mn to 10 MN
- $10 to 50 Mn
- $ 50 to 100 Mn
- $ 100 Mn & Above

- All companies represent the chemical industry in India. Location of respondents are as follows:

  ○ North India

  ○ West India

  ○ East India

  ○ South India

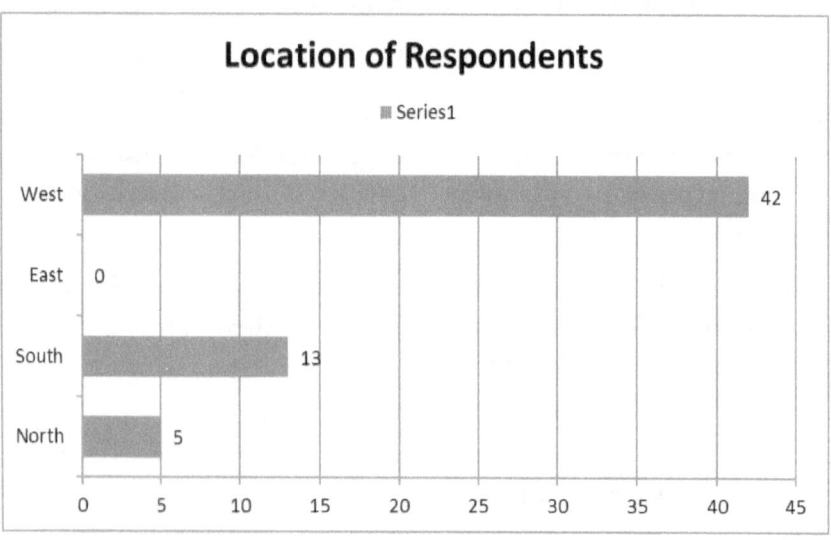

- Based on a survey with 61 chemical companies in India, 43% companies agreed that they add 0 to 5 number of new suppliers every year, whereas for 20% companies, it was more than 15 number of new suppliers added per year, which made "Supplier Selection" criteria and process very critical for the business.

Number of New Suppliers Added per Year	The Response of Participants ~ YES
0 to 5	26
5 to 10	13
10 to 15	10
15 and Above	12
	61

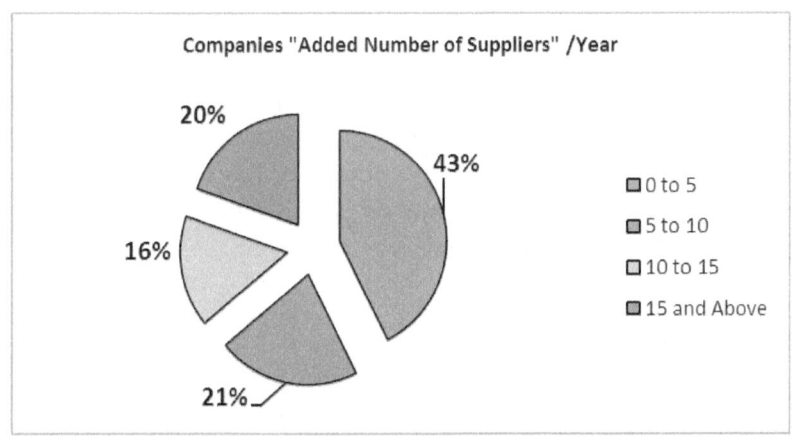

- Based on a survey with 61 chemical companies in India, 10% of companies agreed that they add over 15 new products every year and each new product might require few new suppliers. Whereas, for 62% companies it was 0 to 5 new products added per year, which meant, once again, support was needed for regular new supplier assessment programme within the organisation.

Numbers of New Products Added per Year	The Response of Participants ~ YES
0 to 5	38
5 to 10	13
10 to 15	4
15 and Above	6
	61

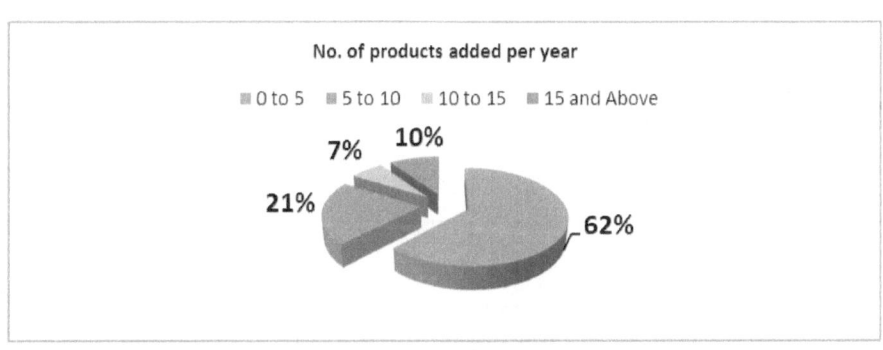

- Based on the survey with 61 chemical companies in India only 6% of companies agreed that they delete or discontinue vendors on a yearly basis. In fact, only 3% of companies agreed that they discontinue 5 to 10 vendors and other 3% agreed that they discontinue 15 and more vendors every year based on their quality assessment, etc. which strengthen the fact that robust supplier assessment process is required as over 90% are engaged on a regular basis by the customer.

Number of Suppliers Deleted/ Discontinued per Year	Response of Participants ~ YES
0 to 5	57
5 to 10	2
10 to 15	0
15 and Above	2

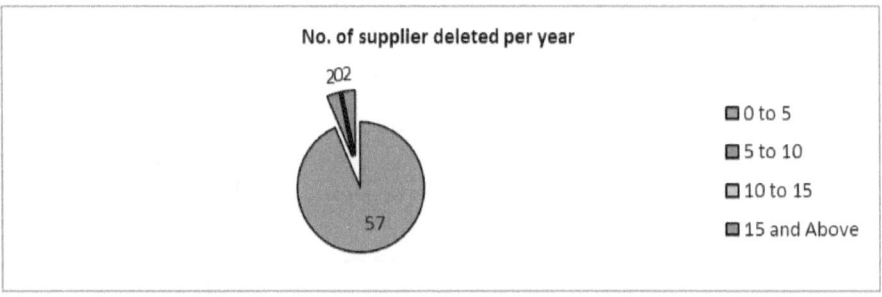

- Based on the survey with 61 chemical companies in India, approx. 60% companies claimed that they conduct an assessment for up to 5 new suppliers on a yearly basis, largely for high-value strategic suppliers; whereas 20% companies conduct an assessment for 5 to 10 suppliers every year and 20% companies agreed that they conduct an assessment for over 10 suppliers every year.

Number of Audits Conducted per Year	Response of Participants ~ YES
0 to 5	36
5 to 10	12
10 to 15	7
15 and Above	6
	61

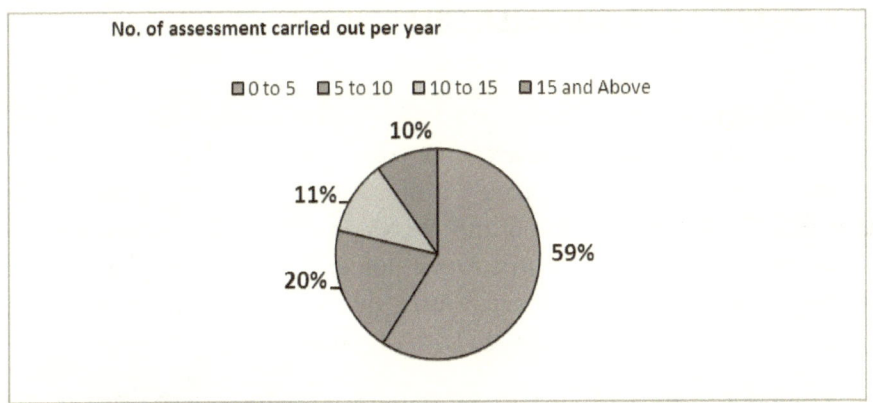

Most of the suppliers selected for this study have, at some point of time in the last 20 years, had an experience of working with the author, therefore, largely, the respondent's category was either from senior level management or from their buying group to ensure the authenticity of data and responses. The author has personally spoken to all the respondents and interviewed them with respect to the core objective of this study, which ensured a high level of reliability. Therefore, while shortlisting suppliers for this study, besides the fact that the author and respondent were known to each other through industry network, the following criteria were kept in mind;

- Can these suppliers deliver information about what is required for this study?

- How long they have been in the business?

- Are they on any approved supplier list from trade associations or government?

- Are they supplying to any overseas customers of export % of their business?

- How frequently they add new products for which they might require a new supplier?

- Have multiple suppliers based on their major raw materials?

- Having domestic as well as overseas supplier base.

- Have core procurement team for supply-chain/supplier development function.

- Consider choosing a supplier as strategic sourcing activity.

- Highly involved or understand how their supplier works.

KEY CRITERIA FOR SUPPLIER SELECTION–
STUDY RESULTS

There are some interesting findings associated with supplier selection criteria and re-confirmed that the concept of cost, quality and on-time delivery are relevant to purchasing managers in the selection of new suppliers today as it was for Dickson in 1966 and Weber in 1991. The survey has also demonstrated that there is a little formal understanding or emphasis on the ability of the supplier to innovate. Firms that value IP and innovation are often reluctant to share information with suppliers.

From the review of current procedures and discussions held during the time of the survey, it is apparent that the vendor selection processes employed in many small to medium-size organisations are ineffective and just a paper exercise or judgement-based decision rather than analysis driven process. In many cases, the selection process is generic and has become a cut and paste exercise taking procedures from processes documented by some other company or organisation. Wherever selection criteria have been applied, they are often not specific to the needs of the organisation or the items being procured. Whereas, in large organisations/ global companies, there is at least some process in place to vendor screening, qualification and selection for new vendors as well as supplier rating system for existing vendors followed by regular audits as per defined time intervals.

While purchasing professionals may be highly trained in their role, they still need support from other functions for an in-depth understanding of all of the needs of other stakeholders. Specific details regarding logistics, specific quality requirements or technical constraints are best communicated by the people directly involved. There is a need for cross-functional cooperation in the identification of the most relevant vendor selection criteria, which is largely considered as procurement department's responsibility and domain to choose and offer the best supplier from the available lot. The selection process and criteria must fit with the current strategy of the organisation. Last but not least, visit the supplier facility by management or selection team to ensure or authenticate assessment details are crucial and very important for the overall success of the programme.

The process of new vendor selection can be very involved and resource intensive. To gain the optimal value, this process may be best applied when selecting suppliers on 80:20 principle, means 20% of those suppliers supplying 80% value are representing 80% spend of the organisation. As a part of the study, they developed the supplier selection template, having 3 parameters, i.e., weight of attributes, supplier rating on 1 to 5 scale and minimum qualification score. Weight of attribute is defined based on survey results and ratings given by management representatives as well as buyers. The given weight is further ratified based on weighed mean average value of the specific segment of attributes given in chart i.e. data collected from 61 vendors on supplier selection process from the vendor's perspective.

For being approved, vendor suppliers need to score minimum qualifying marks as derived in the table next:

Major Attributes of Vendor Assessment	Qualification Score	Min. Score	Max. Score	Mid-Value Point
Suppliers Quality Management System	170	56	280	170
Delivery and Logistics	170	55	275	170
Company Goodwill/Market Reputation	160	51	255	160
Financial Soundness	80	25	125	80
People and Organisation	150	47	235	150
Corp. HS+E and Social Responsibility	120	47	185	120
Manufacturing and Innovation	150	48	240	150
Total Score	**1000**	**329**	**1595**	

Further, in the final template, to address the dynamics of the chemical industry and to keep provision for the importance of various attributes, there is flexibility available in deciding the individual score, whereas, the average score of the specific segment is based on data analysis. For instance, one attribute is "Company Goodwill/Market Reputation", i.e., weighted average value of this attribute was 8.5 out of 10 (1 to 10 scale), since under this attribute there are 6 sub-categories and hence individual weight allotment is kept flexible to decide, based on factors like particular segment of industry in the chemical sector like Pharma, Agro, Textile, Electronics, Polymers for any other chemical intermediates, at the same time average weight is fixed based on survey results, like in this category it is 8.5 whereas weight allotted to sub-categories are $(10 + 7 + 8 + 8 + 9 + 9)/6 = 8.5$

Getting the supply-chain partners wrong can badly affect the efficiency of the organisation, undermining the importance of supplier selection process and requirements might undermine the viability of a company.

Strategic supplier segmentation is a necessary precursor to achieving best practice in supply-chain management. The work identified two different supplier management models: i.e., traditional arm's length approach and collaborative approach means reliable and trusted partners. Further, with respect to preventing supply-chain disruption, one of the most effective ways to manage supply-chain risks is to keep them from happening. Just as Six Sigma companies utilise data and statistical analysis to measure and improve operational performance to prevent quality problems, so companies use data and analysis to significantly reduce the likelihood of supply disruption.

Today's companies that are looking to add a supplier partner or outsource manufacturing to a third party can apply the same standards when deciding where to look for them. Similarly, supplier organisations that follow strict safety standards in their own facilities will choose suppliers that do the same. If customer business is important enough to a supplier, it will allow them for a safety audit of the plant or facility to achieve preferred supplier status. Companies that are truly committed to this process sometimes go so far as to look at the suppliers of their suppliers.

Many companies rush to revamp their supply-chain without giving due focus on the supplier selection process and outsource to China, Hungary, India, Malaysia, Philippines, Vietnam and other developing countries, might need to be careful and properly define their supplier selection criteria to meet their short-term and long-term objectives from a new developed business associate. The message isn't that companies should never outsource to low-developed high exposure territories, but rather that companies can help themselves by factoring the attendant risks into the – decision-making process and weigh those risks against the potential rewards. Where the risks are deemed unacceptable, look for ways to prevent or control them. In short, all these can be only possible when there is a careful assessment done in the beginning while qualifying new supplier.

Vendor selection has attracted researcher's interests for the past decades. However, SMEs vendor selection in India is still an undiscovered area. The sourcing organisation needs to detect its own weakness and strength with threats and opportunities in the back of their minds. The multiple case studies indicate that cost reduction is one of the most important internal driving forces for the companies outsourcing to India.

The multiple case studies further indicate that SMEs simplify the selection criteria and only focus on core factors. To be more specific, product portfolio, corporate culture, support resources, quality systems, internationalisation, negotiability, product performance, customer support, delivery performance, HS+E and cost are emphasised criteria measurements, aspects in SMEs differ from current literature in many aspects.

For instance, most of the case organisations do not check the supplier's quality certificates and registration status, even though it is highly recommended by current literature. Furthermore, most case organisations choose small vendors over large ones, which is contrary to the suggestion given by literature. In the final decision-making phase, SMEs are supposed to make the trade-offs between bargaining power and stability of supply, which reflects on relying on a single source or back up suppliers. Just as the old saying goes, "Do not put all your eggs in one basket," the multiple case study indicates that back up suppliers are more beneficial.

Assessment Format

Given below is self-assessment template for manufacturers to gauge themselves from global buyer perspective and analyse why a foreign buyer should tie-up with their organisation, whether, they offer any specific USP to ensure for a long-term sustainable relationship. An attempt is made to create an example below on the attributes discussed. Weightage means rating decided by most of the foreign buyers for various issues, depending on their criticality to business and factors which can affect the sustainability of association.

Score 5: Strongly agree

Score 4: Agree

Score 3: Satisfactory

Score 2: Disagree

Score 1: Strongly Disagree

Attributes for Assessment	Weight Age	Score Given	Total Points
		1 to 5 Scale	
		(Example)	
Company Goodwill			
Proprietary Information and Protections	10	4	40
Compliance of law and ethical standards	10	4	40
Not a generic competitor	6	3	18
Historical relationship	4	4	16
Value seen in association	4	4	16
Company Vision	4	3	12
People and Organisation			
Health, Safety and Environment	10	5	50
Responsible care	8	3	24
Willingness to change	4	4	16
Human Resource	4	2	8
Leadership commitment	8	4	32
Service Level			
Organisation efficiency and response	6	5	30
Exposure to regulations and doc. Requirement	4	4	16
Financial Flexibility			
Financial Soundness	10	4	40
Cost and Logistics			
Global Cost competitiveness	10	4	40
Supplier reliability	10	3	30
Lead time	8	4	32
Plant locations	4	4	16

Attributes for Assessment	Weight Age	Score Given	Total Points
		1 to 5 Scale	
		(Example)	
Product/Process and Technology			
Manufacturing capacity	10	5	50
Product quality and consistency	10	5	50
Process and facility standards	8	4	32
Technical expertise	8	3	24
Product registrations and handling	8	2	16
R&D/Analytical capability	4	1	04
Technology and engineering support	4	3	12
Contamination and Prevention	4	4	16
			680/900

Note: Minimum qualification points required is 600 for expecting approval from foreign chemical companies for supplies/contract manufacturing and expecting the association to last long.

For important purchases, ensuring acceptable supplier quality is critical. Performing a supplier quality audit is one step in ensuring to the selection of a capable supplier.

A supplier quality audit is more than just walking a supplier's shop floor and looking around. You should leave a supplier quality audit with a greater understanding of at least the following items,

Quality Methodologies – Many quality methodologies are used today including Total Quality Management, Six Sigma and Lean Six Sigma. Ask the supplier about the methodology they employ and how it's used.

Lot Determinations – Organisation defines a lot as "a defined quantity of product accumulated under conditions considered uniform for sampling purposes". In quality recalls, good suppliers know exactly which lots were affected by the non-conforming material, ingredient, etc. So, learn how the supplier tracks lots.

Tools – How is the supplier monitoring its quality? One common tool used is Statistical Process Control (SPC) charts.

SPC identifies an acceptable range of quality measurements and then graphs actual measurements both between upper and lower limits of acceptability and outside those limits. Ask the supplier to show you how quality tools are used to monitor quality.

Improvements – Suppliers often use numbers to gauge their quality performance. Ask the supplier for a few years of reports with these metrics. Are the metrics getting better, worse, more static or more erratic?

Team Empowerment – Experts agree that quality improves when line workers are empowered to watch for defects. Can the supplier's employees tell you what they do when a defect is discovered?

Sampling Arrangements – In mass production, not every unit made is inspected. How does the supplier choose the size and composition of inspected samples?

Inspection Process – How does the supplier inspect items before shipment? Is this inspection method like the way your company evaluates conformance?

The need and importance of the above-discussed attributes vary from organisation to organisation, based on their level of global sourcing, level of outsourcing from one country, organisation maturity and flexibility to adjust with a different culture. The solution in this area has matured greatly over the past few years and have demonstrated proven ability to deliver COSTICIDE* (a word coined to kill cost) to global organisation opting for global sourcing as a cost improvement strategy. It is likely that the thoughts shared above in context to the assessment of source will identify the needs wherever improvements are required prior to the start of commercial transactions. As a result, the effectiveness of a company global sourcing will be an important determinant for the corporation overall success.

Choosing the right supplier involves much more than scanning a series of price lists as choice depends on a wide range of factors such as value for money, quality, reliability and service. The vendor selection process can be very complicated and emotionally undertaking if you don't know how to approach it from the very start. As a buyer, it requires guidance on how to analyse business requirements, search for prospective vendors, lead the team in selecting a right supplier and provide with insight on contract negotiations and avoiding negotiation mistakes. As the scope is limited to right vendor/supplier selection, before we start the following are main steps:

- Gather data or perform an interview to decide on the evaluation team.

- Define major criteria what we are searching in vendor as selection merits.

- Give weight to each criterion and decide on the range for allotting scores.

- Compile a list of possible vendors.

- Select vendors to request for more information.

- Write a request for information with Survey forms.

- Discussion with shortlisted vendors to ensure they have understood the requirements.

- Evaluate response and tabulate the same in a consolidated way.

- Draw a conclusion and propose a solution if any.

- Explore if there can be any future extension of this study w.r.t. specific purpose.

The types of research strategies are usually divided into five different categories: Experimental, Survey, Archival Analysis, Historical Analysis and Case Study. When deciding what type of strategy to choose, the easiest way is to evaluate the possibilities through three different conditions are,

a. Type of research questions

b. The extent of control over behavioural events

c. The degree of contemporary events

The survey strategy was chosen over a case study to enable the use of a larger sample and achieve higher generalizability. It also emphasises on the use of quantitative data followed by statistical analysis, which is in line with the formulation of the research survey.

As a research strategy sample selection of participants based on 3 sizes of companies, i.e., small-scale/medium scale and large scale. As vendor selection is one of the most important decisions for companies. Its strategic nature is both complex and critical to the sourcing company. Analysing different suppliers entails large extents of complexity and uncertainty. This is mainly derived from intangible aspects of relationship and performance factors. Hence, it is necessary to base the vendor selection decision on right criteria associated to specific industry or organisation requirements to fulfil customer expectations, primarily company's motives are dominated by competitive price/cost, quality, lead time, technology and high safety, health and environment standards.

Therefore, the sourcing company aims to select the vendors on the most efficient criteria unique to the situation. By choosing the right criteria, it provides the sourcing company with competitive advantages. Strategic decision-making further differs between MNCs and SMEs in regard to accessibility of internal resources. MNCs plentiful resources support the collection of information, processing and to perform sophisticated interpretations. In contrast, SMEs normally don't access these resources to the same extent which leads to less comprehensive strategic decision-making. Hence, decision makers in SMEs more commonly base decisions on their own cognitive biases and on experiences and lack methodical analysis. The process in which the supplier is selected further has a large influence on the continuous relationship. Overall, supplier selection is a difficult job. A supplier may fulfil certain criteria, but when analysed deeper they fail on other criteria.

As compared to global standards, the Indian chemical industry is still fragmented with a mix of large, medium and small-scale firms operating within the sector. The fiscal concessions granted by the Government to the small sector in the last decade led to the establishment of a large number of – small-scale units in the industry. The sector is divided into three segments namely, basic chemicals, speciality chemicals and the high-end knowledge segment.

The Basic Chemicals segment is the oldest in the Indian chemical industry. It is divided into sub-segments including petrochemicals, inorganic chemicals, organic chemicals, fertilisers and pesticides among others. The speciality chemicals segment is mainly comprised of many small and medium-size manufacturing units.

Over the last decade, the Indian chemical industry has diversified significantly. Today, India has a strong presence in the production of basic organic and inorganic chemicals, pesticides, paints, dyestuffs and intermediates, petrochemicals, fine and speciality chemicals as well as cosmetic and toiletry segments. Use of advanced technology, continuous research and development, development of the domestic capacity to reduce dependence on imports of raw materials are likely to remain key success factors for Indian chemical industry going forward. Besides safety, health and environment protection issues are becoming increasingly important today. Moreover, as more Indian chemical manufacturers expand beyond national boundaries, the need to achieve global standards by improving productivity through better raw material utilisation, bi-product reduction and use, energy reduction and conservation, effluent management, water management, up-gradation of plant and equipment and skill development remains imperative.

Therefore, we have selected mix population from all sizes of companies to reach to a conclusion with respect to "Supplier Selection Process" in Indian chemical industry wherein, handpicked diversified 61 chemical companies from different background and parts of India for primary research, collected information w.r.t. nature of business, how long they have been into business (the fact that almost all the respondents are into manufacturing, and supplier selection is a critical aspect for their business), number of employees, number of products introduced per year, number of new supplier's developed and number of suppliers black-listed or discontinued every year.

A strategic approach to choosing suppliers also helps in understanding how our own potential customers weigh up their purchasing decisions. The most effective suppliers are those who offer products or services that match or exceed the needs of the business. So, while looking for suppliers, it is best to be sure of business needs and what business wants to achieve by buying, rather than simply paying for what suppliers want to sell.

Wherever possible, it is always a good idea to meet a potential supplier face-to-face and see how their business operates. Understanding how supplier works will give an always better sense of how it can benefit business based on this understanding it is ensured that all the respondents are consulted on a personal basis for participating in this survey and tried to understand their process and systems in context to vendor selection.

One can find suppliers through a variety of channels, but it is best to build up a shortlist of possible suppliers through a combination of sources to give a broader base of the selected market, and the same principle is used herein selecting 61 companies chosen for participation in this survey.

As there is a right mix of small–medium–large size suppliers with a presence in domestic as well as international markets, most of the suppliers export 10% to 80% of their products to overseas market. Hence, it is equally important for them to know what are the attributes their customers give due weight to while selecting suppliers.

I spent over 16 years with major US-based large global chemicals and polymer manufacturing organisations (E.I. DuPont & Honeywell). Both the companies hired me to set-up their sourcing function as well as to set up a global sourcing base in India. During these 16 years in chemical industry, I visited over 300+ chemical manufacturing company in India, China, Taiwan, Thailand, USA, Germany, Austria, Czech Republic, Middle East, etc., while the prime focus has been to identify companies having HS+E and other manufacturing infrastructure at par with existing approved global source.

I have visited and inspected over 200 chemical producers within India but was able to qualify only 30% of suppliers/manufacturer and have to extend a lot of handholding to bring them to the level of global standard to grow together. In order to make this job a little easier after a lot of research and understanding defined criteria for supplier assessment and qualification, in a process perfected the assessment template and related scorecard in a way that companies which are able to score above 70% points are really had capability to become global source, eventually result of that assessment process was fantastic, as the companies shortlisted for business for these multinational giants were not only able to develop very complex molecules which were never before produced in India, but in a few cases at a price level almost 5 times cheaper than Europe/US and 2 times cheaper than China, that enhanced the organisation belief system that what we need to do while choosing the "Best Cost Country" for sourcing. Recall those days wherein, I developed a mechanism with respect to vendor plant assessment. I used to visit first their R&D or Quality Control Laboratory lab. The chemical plant is like the cockpit of the aeroplane. Means If cockpit of the aeroplane is in order and follows SOP's and other required HS+E provisions, rules apply to chemical plants as well, there are fairly good chances that if QA laboratory is in order, the plant would be replicating the same culture. Finally, it's organisation culture and commitment which defines their growth.

Supplier Qualification – A Risk Assessment Tool

Supplier Qualification is a crucial practice for any company. However, the practice is applied only partially and not always at the right time. Supplier qualification process needs to assign to each supplier a risk class that depends on the type of product or service to be provided and on the company's risk analysis assessment.

It involves evaluating supplier's compliance with a set of qualification criteria defined at the company level. This evaluation leads to a qualification score and to a qualification validity period. It is carried out by analysing the answers to a questionnaire or assessment sheet which is sent to the supplier and if required by the level of risk by completing this analysis through a detailed audit on supplier's site.

Identify the criteria for which the level of compliance (i.e., the qualification score) needs to be improved and defined for each identified criterion, the improvement action, monitor/control the implementation of the improvement plan. Often, the qualification process is driven by the function or business unit that identifies the need for an alternative source of supply and different functions, business units or teams may conduct their qualifications in different ways.

Companies are recognising that a more robust approach to supplier qualification can have an important role in their product development, risk management and strategic procurement and supplier sustainability aspect. The qualification process encompasses the end-to-end qualification process from supplier and material selection through lab trials, testing and establishing various parameters, control phase 3 supplies at least to ensure plant and fitness-for-purpose, to final sign-off, if everything is in line as per desired results. Tools like Six Sigma are quite helpful at this stage when we are acid testing supplier capabilities to add them in the list of approved supplier list.

Streamlining material qualification processes in this way delivers several benefits for companies, beyond reducing supplier concentration risk and capturing sourcing savings in their largest and most critical material categories. For example, more cost-effective qualification processes allow companies to extend their raw material sourcing cost reduction efforts to categories that would have been too expensive to evaluate previously. Improved robustness of these processes helps to ensure that companies don't run into problems downstream when a new supplier fails to meet the requirements.

By the time we complete this stage, the following must be checked in the organisation:

- Strategic onboarding of supplier.

- Track due diligence activities.

- Document management.

- Regulatory and Compliance requirement.

- Successful completion of plant trial activities.

- Sign-off of Quality and Service agreement.

- Risk assessment and Report finalisation.

SIX SIGMA IN PROCUREMENT WORLD

Using examples of successful implementation of the Six Sigma breakthrough strategy at companies like GE, DuPont, Honeywell and Motorola laid out in detail the theory and practice of achieving the astounding level of quality. The Six Sigma breakthrough strategy is compelling and successful because it focuses on business processes and the components that comprise those processes.

WHY SIX SIGMA

What Drives Companies to Implement Six Sigma?

Six Sigma Is About Improving Profitability: Quality alone is not the most important motivating factor. Each sigma shift provides a 10% net income improvement, a 20% margin improvement and 10% to 30% capital reduction.

Six Sigma Sets Different Standards: Past definitions of a quality focused on conformance to standards as companies strived to create products and services that fell within certain specifications limits.

Six Sigma Is Process-Oriented: Companies use thousands of processes to create their products and services. Six Sigma creates specific improvement goals for every process within an organisation, allowing companies to understand and incorporate new technologies for improved process performance.

Six Sigma Stands for Quality: For some companies, the cost to deliver a quality product can account for as much as 40% of the sales price. No wonder former Motorola CEO Bob Galvin once quoted that leaders must take quality to a personal level in order to create lasting improvements.

The Cost of Quality

Sigma Level	Defects per Million Opportunities	Cost of Quality
Two (2)	308,537	Not applicable
Three (3)	66,807	25% to 40%
Four (4)	6,210	15% to 25%
Five (5)	233	5% to 15%
Six (6)	3.4	< 1%

Each Sigma shift provides 10% income improvement.

1. It is a business process that allows companies to drastically improve their bottom line by designing and monitoring everyday business activities in ways that minimise waste and resources while increasing customer satisfaction.

2. It guides companies into making fewer mistakes in everything they do, from making of purchase orders to the manufacturing of product of Nano-technology, eliminating lapses in quality at the earliest possible occurrence.

3. It does not merely detect and correct errors; it provides specific methods to recreate processes so that errors never arise in the first place.

The Six Sigma Breakthrough Strategy

There are five main phases involved in applying the breakthrough strategy to achieve Six Sigma performance in a process or a company. Each phase is designed to ensure the methodical and disciplined application of the strategy, the correct definition and execution of Six Sigma projects and the incorporation of results in day-to-day business endeavours.

- Define

- Measure

- Analyse

- Improve

- Control

Define: This is also known as identifying and recognising the problem phase, in which companies begin to understand the fundamental concepts of Six Sigma. The key component for companies to address in this phase is variation across process as to how much of an impact, the variation has on results in terms of cost, cycle time and defect rates. The key questions to answer in this phase are; Key Deliverables are Project CTQ, Project Charter and High-Level Process Map.

1. Who is the customer?

2. What is important to the customer–VOC?

3. How does the VOC translate into our language–CTQs?

4. What are the related business goals?

5. What defect am I trying to reduce?

6. Does the problem need to be contained until a permanent fix is implemented?

7. What is the problem?

8. What is the objective of the project?

9. What is the scope of the project?

10. Who are the team members?

11. What is the cost of the defect/benefit for reducing?

12. What is the overall process the project will work on?

13. Where does the process start and finish?

14. How does process output relate to CTQ's?

KEY TOOLS: *Tree Diagram, QFD, AHP, Kano Analysis, Project Charter, Team meetings, Stakeholder analysis, Sigma Track, SIPOC map, Detailed process map, COPQ Estimation, Project Y selection worksheet.*

Measure: It entails breaking down every product into its key characteristics, creating a detailed description of every step in a process and measuring short and long-term process capabilities. The key questions to answer in this phase are: Key deliverables are Project Y, Performance Standards, project data collection and measurement system, data for project Y, Process of the capability of project Y, Improvement goal.

1. What is the goal of the project?

2. How does it relate to business goal?

3. Is the goal is within the team span of influence?

4. Is the goal measurable?

5. What are the performance standards?

6. What are the specification limits and the target?

7. Define the defect, unit and opportunity.

8. What date do I need for my goal?

9. Is my ability to measure good enough?

10. Is the measurement system capable?

11. Is the date acquisition plan in place?

12. Have the data collectors been trained?

13. Has enough data been collected for analysis?

14. Have I established the baseline?

15. Is the date continuous or discrete?

16. What is the current process performance?

17. What is the process capability?

18. What is the improvement goal that satisfies the customer?

KEY TOOLS: QFD, Project Y, Performance standard, Data collection plan, calibration study, Gauge R&R, Historical data, Run chart for a baseline, process capability, Z-Score, Graphical analysis. Benchmarking, Customer specifications, Boxplot, Pareto, Run Chart, etc.

Analyse: It considers where a process is at the time it is measured and points to the goals to which a company should aspire by establishing baseline and benchmarks—thus providing a starting point for measuring improvements. Leadership creates an action plan to close the gap between current and desired processes in order to meet goals for products or services. The key questions to answer in the phase are, Key deliverables have prioritised a list of all X's, List of vital few X's, Quantified financial opportunity.

1. What are the sources of variation for a goal?

2. Have I captured and prioritised all possible reasons?

3. Any quick fix available?

4. What are the vital few reasons?

5. How do you know?

6. What are the test strategies to relate vital reasons to goal?

7. What are the strength and direction of Reason – Goal relationship?

8. What confidence do you have in your conclusion?

9. What is the financial stake?

10. Should the project continue?

KEY TOOLS: Team Brainstorming, Fishbone diagram, Cause and effect, FMEA, Hypothesis testing, ANOVA, Linear regression, Chi-Square, DOE, FMEA, QFD.

Improve: This identifies the steps required to improve a process and reduce the major sources of variation. Key process variables are identified through statistically designed experiments, and the vital few that have the greatest impact are isolated. The knowledge gained from these steps is then used to improve a process, ultimately improving profitability, customer satisfaction and shareholders value. The key questions to answer in this phase are the key deliverables which are, Proposed solution and Piloted solution.

1. What is the relationship between goal and reason?

2. Are reasons operating parameters or critical elements?

3. What is the improvement strategy?

4. What is the right setting for vital reasons?

5. What is the accepted tolerance range for vital reasons?

6. What is the best process flow?

7. Has potential adverse process impact been considered?

8. Let's pilot and confirm experimental results.

9. What are the improvement results of the pilot?

10. What is the potential new process capability?

KEY TOOLS: DOE Design of Experiment, Risk Assessment, Criteria based Decision Matrix, Pugh matrix, Pilot Run, Process Capability of the Piloted Solution,

Control: This phase comprises the integration of Six Sigma into the way a business is managed on a day-to-day basis. More than just a focus of projects through to completion, this stage offers a way to step back and look at how collective results of smaller projects affect the large, high-level processes that run the day-to-day business. The key questions to answer in this phase are the key deliverables like–Sustained solution, Project documentation, Translation opportunities.

1. How will the gain be maintained?

2. Has the control plan been developed and accepted?

3. What is the new process capability post implementation?

4. Can a control chart be used to monitor the new process?

5. Is there a statistical difference between before and after?

6. Has the project final report been completed?

7. Are there leverage opportunities for other projects?

KEY TOOLS: Control plan, Statistical Process Control, SOP, Control Charts, Project closing report, Documentation checklist, Sigma Track, Translation plans.

Levels of Breakthrough; Business, Operations and Process

Almost every organisation can be broken down into three basic levels. The highest level is Business level–the umbrella level that encompasses everything related to the company. The next level is the Operation level, while the lowest is the Process level. The success of Six Sigma is defined as the extent to which it transforms each level of an organisation to improve that organisation overall quality and profitability. The fluidity of the methodology allows it to work up and down the different levels of the organisation.

The Business level application focuses on making a significant improvement to the informational and economic systems used to steer your business, such as customer feedback or supplier quality. It requires 3 to 5-year commitment from executive leadership to consistently do the following:

- Recognise the true state of your business. You can't improve what you do not measure.

- Define what plans must be in place to achieve a higher level of performance and relate that to customer satisfaction.

- Measure the business systems that support the plans.

- Diagnose capability measures and assess performance gaps through analysing benchmarks.

- Improve system elements to achieve performance goals by prioritising efforts for improvement.

- Monitor those efforts and their elements over a period of time.

- Standardise the systems that prove to be best in class.

- Integrate best in class systems into the strategic planning framework.

The operational level issues need to be broken apart into components allowing to define problems, formulate plans and taking positive actions. Since Sourcing is a significant part of Operations, related efforts could be as follows:

- Recognise sourcing issues that link to key business systems, while specific problems can be fixed defects continue to appear sporadically. The company will not be able to improve quality until it has identified the systemic problem.

- Choose projects based on business need, i.e., save cost, dilute sole-source situation, etc.

- Quantitatively gauge how well projects progress, in both an absolute and a relative sense.

- Analyse project performance in relation to operational/sourcing goal.

- Institute regular audits of the Project Management System, ensuring standards that are established and consistently met.

- Standardise best in class management system practices.

- Integrate standardised Six Sigma practices into policies and procedures and reinforce them through rewards and recognition.

In-process level focus is on poor processes that result in problems, additional costs and eroded quality. A disciplined, methodical way of doing things at all level ensures leveraged success. It requires to,

- Recognise functional problems that link to operational issues such as customer satisfaction, profitability, etc.

- Define processes that contribute to the functional problem and effectively search for solutions to problems.

- Measure the capability of each process that offers operational leverage.

- Determine the relationship between the variable factors in the process and determine the direction of improvements.

- Improve the key products/service characteristics created by key processes.

- Control the process variables that exert undue influence.

- Standardise the methods and process that produce best in class performance and integrate them in the design cycle.

- Don't create a new process for every new design or evolution in an existing design.

The approach that one company takes in this endeavour might differ from the approach another takes, but one component is constant. All implementation and deployment strategies must flow down from the leadership. Six Sigma is not a grass-roots initiative. It needs to address factors like dependencies, focus, structure and project selection. Successful implementation depends on several principals;

- Active, Visible, Top-Down leadership.

- Metrics that accurately track the progress of the initiatives, weaving accountability throughout and providing a tangible picture of the company efforts.

- Internal and external benchmarking that provides an honest assessment of the organisation true market position.

- Stretch goals that focus on significant improvement.

How to Select Six Sigma Projects Insourcing

How a company decides to focus its Six Sigma projects directly influences the way Six Sigma is deployed. Companies can focus their efforts on any number of factors, but primarily Six Sigma projects in Sourcing should address the following;

- Cost Savings to boost financial success.

- Sole-Source dilution.

- Verifiable reduction in fixed cost.

- Improve the cost of poor quality, i.e., produce 100% quality the first time through.

- Enhance manufacturing capacity of the organisation either through outsourcing or by creating channel partner.

- Improve cycle time to produce goods or services.

The key to good project selection is to identify and improve those efforts that will boost company financial muscles and impact its customer base positively. It is said that single dollar saving in purchasing adds 100% to organisation profitability.

MEASURING SUPPLIER PERFORMANCE
IS CRITICAL-TO-QUALITY

In global sourcing scenario, sourcing in low-cost countries is all very well, but if the business relies on the dependability of suppliers, then the suppliers need to perform well. Despite supplier performance being an area of concern, over 86% of respondents said that proper and robust supplier selection process can remove these concerns; 48% respondents rated sourcing in low-cost countries as 'Good' whereas, as per 12% respondents it is 'Excellent'; 27% respondents seem to be satisfied with low-cost country supplier performance and rated it as 'Adequate'.

Only 13% of respondents categorised low-cost country suppliers under 'Room for Improvement'. A majority of the respondents confirmed having two or more vendors for each supply, one of which may dominate the other in terms of business share and performance. Vendor rating programmes involve both operational and financial criteria to avoid adverse selection as well as a exercise supplier development programme that can produce tangible benefits such as greater quality and flexibility and more reliable deliveries. In emerging markets, there are constantly new suppliers available, so it is important to continue scouting seeking for the best opportunities and be ready to switch supplier to create new performance benchmark in supply-chain activities.

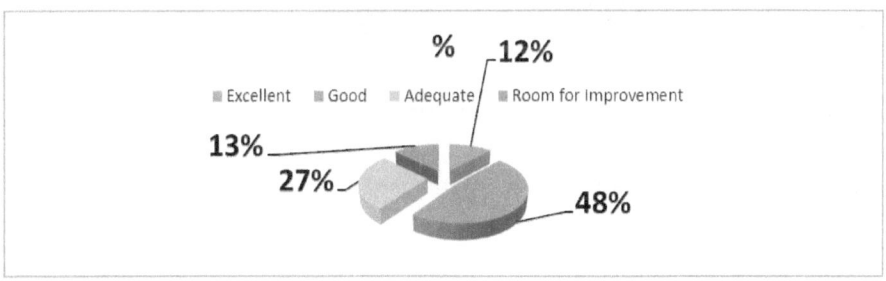

The above data confirms that there are many problematic issues that companies must tackle in order to get the best out of their sourcing operations through systematic Supplier Selection and Development plan.

While evaluating suppliers, as a customer we tend to focus more on supplier resources as the same is highly visible during visit to supplier plants whereas our supplier selection process should have capability to gauge supplier's ability to turn these resources i.e. their physical and human assets such as physical facilities, technologies, tools and workforce into capabilities that in turn can be combined to create high-level customer-facing competencies.

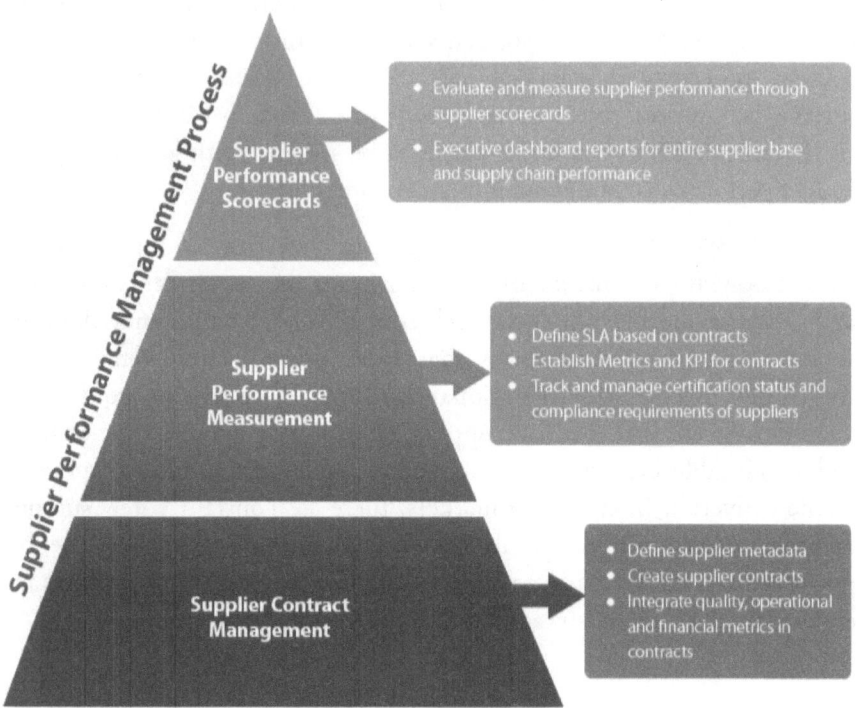

Supplier performance management process deals with the assessment of existing or prospective suppliers on the basis of their delivery, prices, production capacity, quality of management, technical capabilities and services to meet customer requirements and expectations. There are various ways and means to ensure supplier performance like integrating quality, operational and financial metrics in the purchase contract. Track and manage certification status and compliance requirements of suppliers, evaluate and measure supplier performance through supplier scorecards, executive dashboard reports for entire supplier base and supply-chain performance.

While working on supplier performance measurement tool, it is worthwhile to define and develop service level agreement based on the preview of the main contract towards the quality of product and services, packaging, lead time and responsiveness. Some of these complexities in day-to-day business routine work get reduced if due care is taken during the selection of new supplier and all the related necessary attributes are cross-examined whether based on plant visits of supplier, documentation audit, etc.

The supplier selection problem is of vital importance for the operation of every firm because the solution to this problem can directly and substantially affect costs and quality. Therefore, effective supplier evaluation and purchasing processes are critical success factors. A great deal of research is required to determine what criteria should be used to evaluate suppliers pre-commercialisation and with a limited level of information. In practice, any set of criteria can be considered in light of real-life constraints which are specific to particular industry scenario and are making the supplier selection a complicated decision problem that involves balancing many trade-offs.

The chemical industry is one example where the outsourcing of manufacturing is a choice. It is appropriate in some instances but might not in others. For instance, a large bulk chemical producer might reject the outsourcing of manufacturing as an option as it continues to invest in a world-class asset base focused on the global markets and its customers. On the other hand, another company might like to outsource synthesis of chemical intermediates which are critical in the development of final complex molecules and avoid huge investments in setting up additional reaction capabilities if the same is not required for their larger product basket.

As both situations represent a business opportunity within the industry, the company's decision regarding the type of businesses within which it wishes to compete, namely the molecule it will produce and their application and volume, dictates the applicability of outsourcing as a strategic option. On the other hand, competitive pressures and the ongoing search for increased efficiencies and greater profits are driving the industry to re-evaluate their supply-base and relationships to improve overall business performance.

STRATEGY TO REPRESENT INDUSTRY MIX

Choosing the right supplier involves much more than scanning a series of price lists as the choice depends on a wide range of factors such as value for money, quality, reliability and service. The vendor selection process can be very complicated and emotionally undertaking if you don't know how to approach it from the very start. As a buyer, it requires guidance on how to analyse business requirements, search for prospective vendors, lead the team in selecting a right supplier and provide with an insight on contract negotiations and avoiding negotiation mistakes. As the scope of this study is limited to right vendor/supplier selection before we start the following are main steps:

- Gather data or perform an interview to decide on the evaluation team.

- Define major criteria what we are searching in vendor as selection merits.

- Give weight to each criterion and decide on the range for allotting scores.

- Compile a list of possible vendors.

- Select vendors to request for more information.

- Write a request for information with survey forms.

- Discussion with shortlisted vendors to ensure they have understood the requirements.

- Evaluate response and tabulate the same in a consolidated way.

- Draw a conclusion and propose a solution if any.

- Explore if there can be any future extension of this study w.r.t. specific purpose.

The types of study related strategies are usually divided into five different categories: Experimental, Survey, Archival analysis, Historical analysis and case study. When deciding what type of strategy to choose the easiest way is to evaluate the possibilities through three different conditions;

1. Type of research questions

2. Extent of control over behavioural events

3. Degree of contemporary events

As a research strategy sample, selection of participants based on 3 sizes of companies, i.e., Small-scale, Medium scale and Large scale. As vendor selection is one of the most important decisions for companies, its strategic nature is both complex and critical to the sourcing company. Analysing different suppliers entails large extents of complexity and uncertainty. This is mainly derived from intangible aspects of relationship and performance factors. Hence, it is necessary to base the vendor selection decision on right criteria associated to specific industry or organisation requirements to fulfil customer expectations, primarily company's motives are dominated by competitive price/cost, quality, lead time, technology and high safety, health and environment standards. Therefore, the sourcing company aims to select the vendors on the most efficient criteria unique to the situation.

By choosing the right criteria, it provides the sourcing company with competitive advantages. Strategic decision-making further differs between MNCs and SMEs regarding accessibility of internal resources. MNCs plentiful resources support the collection of information, processing and to perform sophisticated interpretations. In contrast, SMEs normally don't access these resources to the same extent which leads to less comprehensive strategic decision-making. Hence, decision makers in SMEs more commonly base decisions on their own cognitive biases and on experiences and lack methodical analysis. The process in which the supplier is selected further has a large influence on the continuous relationship. Overall, supplier selection is a difficult job, a supplier may fulfil certain criteria, but when analysed deeper they fail on other criteria.

As compared to global standards, the Indian chemical industry is still fragmented with a mix of large, medium and small-scale firms operating within the sector. The fiscal concessions granted by the Government to the small sector in the last decade led to the establishment of a large number of – small-scale units in the industry. The sector is divided into three segments namely, basic chemicals, speciality chemicals and the high-end knowledge segment.

Over the last decade, the Indian chemical industry has diversified significantly. Today, India has a strong presence in the production of basic organic and inorganic chemicals, pesticides, paints, dyestuffs and intermediates, petrochemicals, fine and speciality chemicals as well as cosmetic and toiletry segments. Use of advanced technology, continuous research and development, development of the domestic capacity to reduce dependence on imports of raw materials are likely to remain key success factors for Indian chemical industry going forward. Besides, safety, health and environmental protection issues are becoming increasingly important today. Moreover, as more Indian chemical manufacturers expand beyond national boundaries, the need to achieve global standards by improving productivity through better raw material utilisation, bi-product reduction and use, energy reduction and conservation, effluent management, water management, up-gradation of plant and equipment and skill development remains imperative.

Therefore, we have selected a mixed population from all sizes of companies to reach a conclusion with respect to supplier selection process in Indian chemical industry w.r.t. nature of business, how long they have been into business (considering the fact that almost all the respondents are into chemical manufacturing and supplier selection is a critical aspect for their business), number of employees, number of products introduced per year, number of new supplier's developed and number suppliers black-listed or discontinued every year.

A strategic approach to choosing suppliers also helps in understanding how our own potential customers weigh up their purchasing decisions. The most effective suppliers are those who offer products or services that match or exceed the needs of the business. So, while looking for suppliers, it is best to be sure of business needs and what business wants to achieve by buying, rather than simply paying for what suppliers want to sell.

Wherever possible, it is always a good idea to meet a potential supplier face-to-face and see how their business operates. Understanding how supplier works will always give a better sense of how it can benefit business. Based on this understanding it is ensured that all the respondents are consulted on a personal basis for participating in this survey and tried to understand their process and systems in place for vendor selection.

One can find suppliers through a variety of channels, but it is best to build up a shortlist of possible suppliers through a combination of sources to give a broader base of the selected market and the same principle is used herein selecting companies chosen for participation in this survey.

As there is a right mix of small–medium–large size suppliers having their presence into the domestic as well as international market. Most of the supplier exports 10% to 80% of their products to overseas market and hence it is equally important for them to know what attributes their customers give due weight to while selecting suppliers.

FORMAT FOR SUPPLIER ASSESSMENT

Question	Basic Quality Management System	Points (0–5)
1	Is there an Integrated Management System (Quality, Safety, Environment)?	5
2	Are there valid Quality/Environment/Safety Management certificates? If no, is a certification planned? When?	5
3	Are tasks and responsibilities clearly defined? Are there organisation charts? How are delegation and deputyship done?	5
4	The responsibility of top management for Quality, Delivery, Safety and Environment?	5
5	Process Management Manual available?	5
6	Documented Quality policy/Quality goals/Quality key figures How are they defined, communicated, tracked?	5
7	Involvement of employees into the QM System	5
8	How is the effectiveness of QM System verified and communicated?	5
9	Is there a process for internal audits?	5
10	Is there a plan for internal QM/Process-audits? Covering all processes? Main focus?	5
11	Auditors: Number, selection, independence, training	5
12	Follow-up of corrective actions from internal audits (implementation, effectivity)	5
13	Is there a documented procedure for creation and control of quality documents?	5
14	How are procedures and responsibilities for creation and control of quality records defined?	5
15	How is storage time and place for quality documents and records defined?	5
16	Are Test Specification and Test Method available at the time?	5

Question	Basic Quality Management System	Points (0–5)
17	Whether QC and Production are independent of each other?	5
18	How identification and traceability are established?	5
19	Complaint handling system	5
20	Is the premises situated in an environment which causes minimum risk of contamination to materials and product?	5
21	Is the design and construction for good sanitation?	5
22	Is area provided commensurate with production activities and adequate storage area for under process material provided? Total Area of Factory and area for further product	5
23	Are SOP's available for maintenance, cleaning and sanitation of buildings, premises and surroundings?	5
24	Check whether the area is cleaned before starting processing operations?	5
25	Are arrangements made for control of entry, of rodents, insects, birds, etc.?	5
26	Separate restrooms and refreshments rooms?	5
27	Separate change room and toilets (away from production/storage area) for male and female workers?	5
28	Batch production record docket.	5
29	Reference standard are maintained	5
30	Is there provision for UN approved pkg. wherever required?	5
31	Complaint handling procedures.	5

Question	Quality Assurance	Max. Points
1	Are there inspection plans, written inspection instructions and inspection records? (sample)	5
2	Is test equipment systematically monitored? (Label? Reminder?) Examples?	5
3	Qualification of test methods and test equipment – are they suitable?	5
4	Are there written instructions for release, calibration, identification and monitoring of test equipment?	5
5	Is only test equipment with sufficient accuracy used?	5
6	How are samples taken?	5
7	Is sampling procedures adequate? Do they use test certificate to buy/use?	5
8	Do the label on sample container, show batch no., name and container no., date of sampling and person who sampled. Do they have access to library reference of technical reference?	5
9	Are analytical records and protocols of each batch of the raw materials, intermediaries and finished products available with worksheets?	5
10	Are all retention samples kept till expiry date?	5
11	Is design of lab suitable with adequate space, ventilation and for prevention of fumes?	5
12	Are records of service and calibration on instruments available?	5
13	Are instruments checked daily or prior to use?	5
14	Are date of calibration, service and date when recalibration due indicated?	5
15	Are the reagents/samples prepared according to written procedures and labelled appropriately?	5

Question	Quality Assurance	Max. Points
16	Are references standards available correspond to the materials analysed?	5
17	Are the labels of batch prepared show concentration, standardisation factor, shelf life, standardisation date; storage condition and signature of the persons prepared the reagents?	5
18	Is there a system that assures the capability of test equipment?	5
19	How are test methods validated? Gauge R&R, etc.	5
20	Are all records retained up to expiry period of finished goods?	5

Question	Supplier Rating and – Selection (Suppliers of Supplier)	Points
1	What Quality Requirements are needed for raw materials, Packaging, Service requested from the pre-supplier? How are these requirements made? Are they clear and comprehensive? How are they checked?	5
2	Process selection and release of suppliers?	5
3	"Initial sample testing" performed as part of the release of a supplier?	5
4	Second Source Strategy? Are second sources available for all products? Is supplier qualified? Are materials bought from the second source on a regular basis?	5
5	Does the supplier inform us about changes in the raw material supplier (quality agreement)?	5
6	How is the quality of incoming material assured?	5
7	Is there a system for the evaluation of suppliers? (criteria, consequences)	5
8	How are the QM Systems of suppliers developed?	5
9	Does the supplier have a quality agreement with his suppliers?	5

Question	Supplier Rating and – Selection (Suppliers of Supplier)	Points
10	How is traceability of delivered goods assured?	5
11	Does the supplier perform supplier audits? (qualification, range, consequences)	5
12	Does purchase department staff knowledgeable about the products and materials? Have specifications of product and materials?	5
13	Is the vendor rating system being followed?	5

Question	Inventory Management	Max. Points
1	Are there instructions for handling of chemicals? (dangerous substances)	5
2	How is damage or deterioration of quality during transportation or storage prevented?	5
3	How is the identification of goods during transportation and storage assured? (Labels for single containers)	5
4	Protection against cross-contamination housekeeping in the warehouse/	5
5	What kinds of transportation locks are used?	5
6	How is adequate packaging material determined? – Quality – Transportation – Dangerous goods	5
7	How does the supplier assure that only released material is shipped?	5
8	Segregation of material: released – under inspection – rejected?	5
9	Waivers?	5
10	Days of supply – Assurance of delivery capacity	5
11	Quality of process/key figures (OTTR)	5

Question	Inventory Management	Max. Points
12	How are MSDSs and technical data sheets issued? Are they comprehensible?	5
13	Standard procedure for Source of Raw Materials and acking aterial	5
14	Standard procedure for Receipt of material	5
15	Standard procedure for Goods inspection	5
16	Standard procedure for Truck inspection	5
17	Standard procedure for Goods storage	5
18	Standard procedure for Dispensing of RM/PM	5
19	Standard procedure for Handling of rejected raw and packing materials	5
20	Are batches of starting materials segregated adequately and stocks rotated by FIFO system?	5
21	Is conventional u colour-coding system used to indicate status?	5
22	Is adequate area for finished goods provided?	5
23	Is adequate area for recalled/returned goods provided?	5
24	How the waste materials from the store, production, quality control is disposed off?	5
25	Is area sufficient for storage of materials (Brief description of stores Viz if it is a shed or rooms or a combination of both)?	5
26	Are receiving bays covered?	5
27	Check whether any demarcation is being provided – RM, PM, Approved, Rejected.	5
28	Are records for action taken on rejected goods available?	5
29	Are facilities available for storage in controlled temperature/humidity? *Do companies follow MSDS and Safety Practices?	5
30	Is record of temperature and humidity maintained?	5

Question	Selection of Personnel and Qualification	Max. Points
1	How are new employees selected?	5
2	How are new employees trained?	5
3	Is the performance of employees rated?	5
4	Is there a programme for personnel development?	5
5	How is the training demand identified? Are there training plans?	5
6	How are training done? (Theory/practical)	5
7	A portion of Q-, S- and E-subjects in training?	5
8	How is the effectiveness of training proofed?	5
9	Is there a qualification matrix?	5
10	Matrix for deputies	5
11	How is the motivation of employees increased?	5
12	Involvement of employees regarding Quality, Safety and Environment	5
13	Are employees allowed to stop a process in case of a process, quality, safety or environmental defect?	5
14	Do employees know about the business objectives and current status?	5
15	Medical examination of staff and workers	5
16	The SOP for training available with area/scope of training	5
17	Clothing and Hygiene of personnel.	5

Question	Customer Satisfaction	Max. Points
1	Customer satisfaction evaluation: How? Key figures? Follow-up?	5
2	How is quality agreements negotiated with suppliers?	5
3	How does the supplier assure that customers are informed about changes (kind of changes), MOC?	5
4	The organisation of Customer Service? Defined contact persons?	5
5	Defined process for handling of customer complaints?	5
6	Root cause analysis, the definition of corrective actions, check of effectiveness	5
7	Is there a process for handling a recall? Responsibilities clearly defined?	5
8	Handling of Delivery misses?	5
9	Management of Change or Rapid Problem-Solving process to handle delivery misses?	5
10	Handling of non-conforming products? Documentation?	5
11	Agreement of customer for a special release of non-conforming products necessary?	5

Question	Corrective Actions	Max. Points
1	Are process parameters analysed? e.g. SPC	5
2	Are deviations investigated systematically and corrective actions defined? (learn from mistakes)	5
3	How is the implementation and effectiveness of corrective actions verified?	5
4	Is there a system for identification and avoidance of repetitive mistakes	5
5	Are responsibilities for definition and monitoring of corrective actions clearly defined?	5
6	Documentation of corrective and preventive actions?	5
7	Documentation of Management of Change activities?	5
8	Process for continuous improvement?	5
9	How is the P-, Q-, S- and E-System improved?	5
10	Risk Management. How are potential risks evaluated and how are appropriate preventive actions defined?	5

Question	Security	Max. Points
1	Programme for security	5
2	Is there a security programme?	5
3	A crisis management team is designated and trained?	5
4	Is there a recall procedure?	
5	Is the effectiveness of this programme checked regularly?	5
6	Are there warehouses outside the grounds? Are they part of the safety/security programme?	5
7	Are there written procedures for contractors? (if applicable)	5

	Site	Points
1	Effective measures are taken to restrict unauthorised access to the grounds.	5
2	Surveillance by the camera and/or security guards?	5
3	Effective measures are taken to restrict unauthorised access to sensitive areas.	5
4	Employees and Visitors	5
5	Job applicants and contractors are screened for employment references, residency status, illicit drug use and criminal background.	5
6	Surveillance of access to the facility is done.	5
7	Are employees trained on Food Security?	5
8	Are visitors checked before entering the site?	5
9	Is there a code for behaviours for visitors and contractors? Is it communicated in a written form?	5
10	Are there appropriate garment rules for visitors and contractors?	5
11	Effective measures are taken to restrict unauthorised access to sensitive areas for visitors and contractors	5

	Incoming Goods	Points
1	Is there a procedure for handling and checking of incoming material?	5
2	Is there a process for checking of tank cars?	5
3	Facility Operations	5
4	Are critical areas identified and access restricted?	5
5	What kind of water is used for production and how is it analysed?	5
6	Is traceability assured for all products and packaging materials (even in case of rework)?	5
7	Are the containers tamper-proof?	5
8	Is there a procedure for handling of labels?	5
9	Are microbiological tests performed?	5

S. No.	Production/Filling incl. site tour	Max. Points
1	Are there manufacturing instructions? How are they controlled? Current version available? Example?	5
2	Which capability tests are performed for new equipment?	5
3	Are there batch records? Example?	5
4	Housekeeping; Utility	5
5	Are there instructions and records for cleaning? Example?	5
6	How is equipment released for production? (after cleaning, maintenance)	5
7	Is there a system for identification of product status? (released/not tested/rejected)	5
8	How is implementation of quality service agreements assured? Are the requirements known? When and how are customers informed?	5
9	In-process testing performed?	5
10	Monitoring of measurement devices – How is it organised? – 2–3 examples	5
11	Tracking and evaluation of critical process parameters	5
12	Is supply availability assured in case of process disruption?	5
13	Error prevention? Is there a double check? Product-FMEA?	5
14	Is traceability for products/raw materials assured? Lot numbers…	5
15	Is the maintenance of production equipment done according to a maintenance schedule? Preventive vs. reactive	5
16	Procedure for Change Management (document and record, example)	5
17	How does the supplier deal with unplanned process deviations? (Examples) How is this term defined?	5

S. No.	Production/Filling incl. site tour	Max. Points
18	Which key figures are detected and evaluated?	5
19	Preventive maintenance schedule.	5
20	Equipment maintenance logs.	5
21	Equipment use logs.	5
22	Equipment cleaning procedures.	5
23	Operating Procedures.	5
24	Calibration of balances and measuring devices.	5
25	Sanitisation of pipes/equipment	5
26	Sanitisation of vessels.	5
27	SOP on Production Process and Production Planning	5
28	Is there any deviation from SOPs and manufacturing procedures?	5
29	Mention the procedure adopted for handling/using recoverable rejects.	5
30	Has any recalled/returned batch been converted for resale by relabeling (mention instances if any)?	5
31	Outsourcing, if any	5
32	Is there any provision for dust control in the production area?	5
33	Is the adequate separation of packing lines to prevent any mix-up?	5
34	Is each packaging line identified with product, batch and packaging size?	5
35	Is line cleared before each batch operation, certified and recorded?	5
36	Are all coded packing materials verified before use?	5
37	Is reconciliation of packing materials used as vis-à-vis and products-recorded?	5
38	Is one-line control procedure during packing available?	5
39	Is layout SEQUENTIAL and have labelled on it? Or as per cGMP standard	5

S. No.	Production/Filling incl. site tour	Max. Points
40	Are the production activities conducted under the supervision of expert technical staff?	5
41	Whether the access to production area restricted to authorised personnel?	5
42	Are walls, floors and ceiling free from cracks/dust/cobwebs/spot?	5
43	Is the production area provided with proper ventilation having air control facilities?	5
44	How is cross-contamination of raw material/product by another material/product being prevented?	5
45	Is adequate working space provided for each equipment for easy functioning and cleaning?	5
46	Are there any open channel drains? if yes, are these covered properly? Check the cleaning, procedure and frequency of cleaning.	5
47	Are satisfactory arrangements are being made for storage of change parts and machine tools?	5
48	Are service line painted to indicate the contents and direction of flow?	5
49	Are measuring equipment calibrated periodically?	5

Procurement Performance Governance–An Edge to Sustain Excellence

Managing procurement organisation's performance is an integral component of good governance and improving procurement outcomes. The performance management tool provides a basis for continuous improvement and reporting to internal and external stakeholders. The tool lists a number of measures to report against. Areas highlighted below are suggested as minimum reporting standards for performance management. The greater the number of indicators reported against, the more comprehensive an overview of performance. The tool should be tailored to reflect the specific performance measures applicable to the structure of the organisation and the complexity of your procurement activities. The tool provides two examples that detail how relevant elements of the tool can be synthesised to present performance measures to internal and external stakeholders.

Financial Aspects of Procurement Governance

1. Financial performance of the Top 10 categories/spend.

2. Growth of the Top 10 suppliers by spend.

3. Summary of Procurement Complexity, i.e., spend under transactional, spend under focused spend.

4. Procurement savings against procurement spend/variance plan.

5. Total leveraged/auctioned spend against proposed procurement.

6. Procurement spends by category.

7. Procurement spends by the supplier.

8. Off contract spend.

9. Cost of the procurement function.

10. Year-on-year based cost improvement.

Operational Aspects to Enhance Governance Level

1. Top 5 procurement related risks exist for the organisation, i.e., quality rejections, delivery time, short or excess supply, packaging and logistics and storage aspects.

2. List of short-term and long-term procurement improvement initiatives.

3. Contract management, i.e., expiry, renewal, failure and compliance tracking as well as value targets to be achieved from the contract.

4. Management and resolution aspects towards conflict of interest.

5. Continuous improvement plan w.r.t supplier performance.

6. Risk management and mitigation strategy.

7. Organisation wise procurement plan and strategy.

Somewhat quality of our suppliers, the complexity of transaction and urgency of business requirements affect management and staff aspects. Therefore, procurement staff or people are equally important governance aspect.

1. Number of people in central procurement and regional team.

2. Qualification and experience of procurement staff.

3. Capability assessment of procurement staff.

4. Strategy and structure required to manage business needs.

Dimension	Measurement Aimed at	C/I*	Examples
Purchased materials prices and costs	Purchased materials cost control	C	Materials budgets, variance reports, price inflation, reports, purchasing turnover
	Purchased materials cost reduction	C	Purchasing cost saving and avoidances, impact on return and investment
Product/quality of purchased materials	Early purchasing involvement in design and development	I	Time spent by purchasing on design and engineering projects, sampling reject rate (%)
	Incoming inspection quality control and assurance	C	Reject rate (%), line reject rate (%), quality costs per supplier
Purchasing logistics and supply	Monitoring requisitioning Delivery reliability (quality and quantity)	I/C	Purchasing administration lead times, order backlog (per buyer), rush orders, delivery reliability index per supplier, materials shortages, inventory turnover ratio, JIT deliveries
Purchasing staff and organisation	Training and motivation of purchasing staff Purchasing management quality Purchasing systems and procedures Purchasing research	I	Time and workload analysis of the purchasing department, purchasing budget, purchasing and supply audit

*Note: C = Continuous and I = Incidental

Procurement performance governance thus can be defined as the extent to which the Purchasing function is able to realise its predetermined goals at the sacrifice of a minimum of the company's resources (i.e. costs) which can be measured in terms of effectiveness i.e. results or efficiency at the actual cost.

Further, through a procurement audit, management may assess the extent to which goals and objectives of the purchasing department are balanced with its resources. These audits can be conducted in such a way that people do not feel threatened and in a way that builds trust and generates professionalism. These audits can be preventive or corrective to ensure expectations of excellence path.

MISTAKES DURING MEASURING BUYERS' PERFORMANCE

Measuring buyer performance can be tricky. Here are three mistakes commonly made when setting goals against which buyer performance is measured.

Mistake #1 – *Having Cost Savings Be the only Metric.* One of the most important metrics a purchasing department can share with top management is cost savings. But just because it is one of the most important metrics, doesn't mean it should be the only metric.

If you only measure buyer performance on cost savings, it could incent buyers to sacrifice quality, on-time delivery and/or supplier service for lower prices. So a cost savings metric should be balanced by measuring these other aspects of purchasing performance to produce a clear assessment of the buyer's impact on total cost and overall company performance.

Mistake #2 – *Not Using Net Cost Savings as a Metric.* When calculating cost savings, price increases should be deducted from price reductions to produce a NET cost savings number.

One reason for doing this is that top management expects reported cost savings to equal actual profit improvement.

But also consider buyer motivations. By counting only gross cost savings, buyers may be inclined to ignore opportunities to minimise price increases on large spend categories while focusing their time on less critical categories where price reductions are possible, resulting in an overall lower positive impact on profit.

Mistake #3 – *Not Taking Markets In to Account When Setting Goals.* When some purchasing managers set cost savings goals, they look to last year's numbers or some other arbitrary figure to determine the targets for the next year. This can set buyers up for failure.

In the last few years, there has been cost volatility in many markets. If the purchasing department promises year-over-year price reductions in markets where prices are rapidly rising industry-wide, top management will likely be disappointed when actual performance is compared with those goals. So, give consideration to market conditions when setting buyer goals.

CONTRACT MANAGEMENT PROCESS CAN SECURE PERFORMANCE

Good contract management process ensures that both the buying organisation and the supplier fulfil all of the obligations that they agreed to in signing a contract. It also helps the buying organisation achieve all of the benefits it is expected to when contracting with a supplier.

Not having a good contract management process in place has consequences both to the buying organisation as well as to the purchaser. If no one is managing the performance of the contract, an organisation is likely to fail to meet its goals that it had for the project associated with the contract. Such failures may include delayed timelines, cost overruns and more.

If the purchaser negotiates a great contract and drops the ball in making sure that its terms and conditions were adhered to, s/he may end up looking for a new job quite soon.

Contract process one of the most underrated aspects of purchasing. Here's why…

Lots of attention gets paid to the processes leading up to the signing of a contract. Things like strategic sourcing and negotiation. Those processes get attention because they produce the first statement of how much money will be saved over the contract's term. And that's great.

But the savings numbers that are shared at that point are estimates only. No savings have actually been realised. And they may never be realised. Yet, in many organisations, those same estimates are recorded, the purchasing department takes credit for them and no one ever verifies if those estimates ended up being accurate.

But today's senior executives are getting smarter. If savings are estimated, they want to see exactly how and when they affect the organisation's financial statements.

Without contract management to ensure that those savings are realised, it's likely that the estimates will differ greatly from the actuals. That's when senior executives look someone to hold accountable.

Starting a contract management process doesn't have to be complicated. You can always begin small and grow.

Starting may be as simple as creating a spreadsheet with dates and tasks that must be finished by those dates. The purchaser simply monitors those dates and keeps in close contact with the supplier to ensure timely performance.

For major contracts, the purchaser should also hold regular periodic reviews with suppliers to keep channels of communication open and discuss strategic issues.

When to Involve Supplier

Involving suppliers when developing a new product or service can be smart. But if not done well, early supplier involvement can lead to suppliers taking advantage of customers who lack the future threat of competition to reduce costs and improve performance.

Most companies use the following tactics to control early supplier involvement deals:

1. *Make Cost Data Sharing Mandatory.* For the privilege of their involvement in development, a supplier should be required to break down its quoted price into its component costs and profit margin.

2. *Minimise the Overhead Percentage.* Scrutinise how overhead is calculated. Re-identifying overhead costs as direct costs make it easier to jointly reduce those costs later. For example, costs associated with scrapped material might be buried in overhead. That should be made its own line item.

3. *Understand All Assumptions.* There are always assumptions built into a supplier's cost structure. For example, a supplier may base

its labour costs on an assumed production rate (e.g., 100 units per hour). Document all of these assumptions and have a technical team member evaluate their accuracy.

4. ***Agree to the Right Terms.*** Suppliers who overestimated their costs (or intentionally quoted them higher) should not benefit. So, pricing should be based on a cost-plus fixed fee scheme. Suppliers must agree to share accounting records of their work for you. And you should agree on the terms that can change if the volume exceeds your estimates.

5. ***Audit the Supplier's Books.*** You and your technical team must audit the supplier's records to compare actual costs versus estimated costs and those assumptions you documented earlier with actual results. Where costs were lower than estimated or actual performance was better than assumed, a price adjustment is warranted.

6. ***Continuously Evaluate Cost Reduction Opportunities.*** Auditing supplier's books can also involve evaluating ways to reduce costs. Try to identify materials used by several suppliers to see if consolidating the buy can reduce costs.

MANAGING YOUR SUPPLIER RELATIONSHIP

Are you truly managing your supplier relationships?

Use these tactics for managing supplier relationships:

Supplier Performance Evaluation – Ask a supplier's representative how he thinks the supplier is performing and you may hear "Great!" But what if you think the supplier is performing poorly? Who is right? You can't tell without agreeing upon performance standards. For your strategic suppliers, agreeing upon what to measure (e.g., the percentage of orders delivered by their due date) and what the goal is (e.g., 95% on-time deliveries).

Idea Sourcing and Value Creation – Better profitability can come from ideas. You can greatly increase the number of good ideas by sourcing ideas from your suppliers rather than just from your company's employees. Some leading organisations have systematic processes in place to collect ideas from suppliers, measure their impact and reward suppliers for them.

Supplier Development – It's logical that when you improve the capabilities of your company's workforce, your company benefits. But even though suppliers do the work once done in-house, that logic doesn't always work. Leading companies engage in supplier development – providing resources to improve their suppliers' capabilities. This often involves training suppliers in methodologies such as Six Sigma or Lean, but really can be any collaboration that makes suppliers more capable of delivering benefit to your company.

A Joint Review of Purchase Costs – If you work for a big company, you have a lot of buying power. Buying power that may be wasted if your smaller suppliers have 100% responsibility for buying all the goods and services needed to provide their product or service to you. By jointly reviewing costs further down the supply-chain, you may find opportunities where you can buy some goods and services your suppliers need at a lower cost, ultimately reducing your overall costs.

Building a Better Relationship – To gather all the information necessary to build a strong supplier relationship, you'll need to rely on multiple informal data-gathering processes. Spend time with the supplier. When a supplier looks at who they think will make them successful, they're counting on not only the things you put in writing but also the things you don't have team meetings with suppliers. These periodic reviews bring together top suppliers to discuss the strategic plan. We constantly get input on where we are now and how we need to be poised for the future.

Try to hold regular face-to-face conversations with your key supplier contacts. Face-to-face or more casual conversations are useful with suppliers who are unable to articulate, hesitant to voice concerns informal meetings or feel uneasy about putting issues in writing. Those informal discussions are critical in developing mutual trust relationships are formed at every level, not just among senior managers.

You must develop a relationship from the senior leaders all the way to the purchasing executive who is with the supplier every day. This takes the whole team—the engineers, purchase people, the guys on the shop floor, everyone. In fact, because frontline employees are often the most closely involved with the product or service, their communication tends to be the most accurate and the most valuable.

Team dinners, kick-off meetings, product conventions or sporting events offer great opportunities to get acquainted with suppliers outside the office.

You can really build relationships at these events. Sometimes we get involved in huge deals and forget about the people we're dealing with. Going to these events unites both sides and gets us to work better with each other.

"Better understanding happens when you know more than just the business side of the person you do business with." The core of a successful supplier relationship is what you do with the information you gather. To leverage that data, you have to put it in context. "If you can understand the market a little better, you can anticipate your supplier strategy and develop your own strategy."

Understand what their goals are, what they lose sleep over. That way you can get ahead of the types of solutions they really need. Every relationship has challenges, but many problems can become opportunities to forge even stronger bonds.

"Building relationships is done best when it's problem resolution time." People have a vested interest in the outcome, far more than with easy successes and when the problem is resolved positively, there's more strength and trust built into the relationship. Ultimately, what transforms a relationship with a supplier into a true partnership is sharing the same goals

NEGOTIATION–HOW IT WORKS IN INTERNATIONAL SCENARIO

Negotiation is a continuous process whereby parties with conflicting aims establish terms on which they will cooperate; this process caters to need satisfaction, as both the parties have needs. Therefore, skilled negotiators seek to fully understand other party need and failure to understand above may result into unsuccessful negotiation, as the other party is acting for his or her reason…not yours and hence diagnosing the other party's need will help in suggesting a right strategy.

It is very important that we need to become aware of our personal underlying attitudes and tendencies from which we unconsciously tend to operate. Few of the common negotiation tactics are;

- Gather information for negotiation.

- Make other party tender first offer.

- Suggest hypothetical other party offer.

- Designate a demand as a pre-condition to negotiations.

- Make your major demand at the beginning.

- First and final offer concept.

- Invoke competition.

- Alter other party perception.

- Resist change to your own position.

- Pursue one item while holding back the real item of interest.

- Make a team of two members–one reasonable and non-threatening and other unreasonable and demanding.

Negotiation is all about knowing the targets and then working for the same, the style of negotiators could be different. It can be either competitive negotiator or co-operative negotiator.

Competitive negotiators are more likely to;

- Seek to dominate the other party.

- Prefer to start with tough, often unreasonable demands.

- Tend to be inflexible.

- Demand major concessions, while conceding little.

- Withhold information intentionally and bluff.

- Make a statement, rather than asking questions.

- Use power to obtain compliance, threaten.

- Have solutions pre-conceived from the start.

- Show little or no interest in other party needs.

- Focus on short-term gains.

- Think in terms of win/lose.

Whereas co-operative negotiators are more likely to;

- Interact with other party as equals.

- Regard parties as collaborators.

- Prefer to start with reasonable/realistic demands.

- Tend to be more flexible.

- Be more rational, make use of emotions.

- Be willing to share information, more open and trusting.

- Ask a question, rather making statements–2-way communication.

- Discuss, compromise, rather than the use of power to coerce.

- Show interest in the needs of the other party.

- Focus on long-term gains for both the parties.

- Think in terms of win/win.

One of the reasons that negotiation is one of the most exciting business processes is that there isn't 100% certainty.

We'll talk about three negotiation tactics that have proven to be very effective, yet still can fail if not applied in appropriate situations.

The Crying Poor Tactic: The Crying Poor Tactic is used by publicly-held companies whose poor financial performance is well known, sometimes also used by small companies. Buyers from these companies will stress to the supplier their financial state (e.g., "You know we don't have a lot of money, so we need lower pricing.").

This negotiation tactic can indeed be effective. However, it can also raise suspicion in your supplier that can have a negative effect on the deal. The supplier may be wary that they won't be paid on time or at all. The supplier may worry that your company won't be around to fulfil its contractual commitments. And as a result, the supplier may withhold its best deal.

The Get It from Someone Else Tactic: The fear of losing a deal to a competitor can get a supplier to its lowest price very quickly. Saying "If you can't lower your price, that's OK–we'll buy it from someone else," can work in certain competitive situations.

However, not all markets are so competitive where you can indeed buy the same exact item from any supplier and get equivalent quality, delivery and service. Suppliers in these less competitive markets know this. So, in such markets if, you attempt to use the Get It from Someone Else Tactic on them, they will realise that you have less knowledge of the marketplace than them, which will make them feel like they have more leverage in the negotiation and they will be less likely to concede.

Saving the Toughest Issue for the Last Tactic: Deciding the order in which the various issues will be discussed with your supplier is critical. Some negotiators like to save the toughest issue for last.

Saving the toughest issue for last does work well in certain situations. It allows you and your supplier to agree on easier issues, thus building a rapport and spirit of cooperating for mutual success. It also helps you assess your supplier's negotiation style, strengths and weaknesses while preparing to negotiate the big issues.

However, if there is a deadline for your negotiation, saving the most difficult issue for last can be disastrous. You could have little time left to finish the negotiation and feel pressured to concede to a less-than-optimal deal.

So, while deadlines can work for you in a negotiation, they can also work against you. Evaluate the deadlines: When are they? Who has imposed them? Are they negotiable? And if the deadline has been imposed by an internal customer or management and is not negotiable, address the most difficult issues sooner rather than later.

Effective Negotiator–Playing the Whole Game

The golden rule of negotiation is that "Everything is Negotiable." Don't Use All of Your Ammunition at Once. There may be several logical reasons for a supplier to reduce his price: ordering early, being eligible for a multiple-product – purchase discount, etc. Suppliers will often try to convince you that their first price reduction is a great deal that you should accept. So, if you used all your reasons for getting a price reduction at once, the supplier has more power to defend his price and you have run out of reasons to get a lower price. So follow these steps:

- Know the reasons for requesting a price reduction.
- Decide the sequence to introduce those reasons.
- Ask for a price reduction based on one reason.
- Thank the supplier after getting it.
- Ask for a better price for another reason and repeat.

Sometimes, simply asking for a better price will get you a better deal. At other times, the value of your prospective purchase may be too small to qualify for a lower price. But consequences for simply asking for a better price are rare. So, you should always at least ask, regardless of the likelihood that you'll actually get a better price.

Sellers use psychology-based negotiation techniques to dissuade you from asking for a better price. If you are aware of these techniques, you can resist being influenced by them and you'll always remember to ask for that discount.

Let me share one of the interesting project related negotiation I had been through while negotiating a global deal for large MNC for Agro Chemical i.e. Contract manufacturing of Herbicide with a potential Indian alliance partner. This was a few million dollars project, i.e., to bring one of the Herbicide productions in India, whereas US MNC has an option to take this project to China. Our cost benchmark was lower than our internal cost of producing the same in the USA. It was a case of 3P plus technology transfer. There were team of 7 people from our US team who travelled to India including the CTO, Global Business Head, Global Purchasing Head and other people from various function, there were same number of people from the Indian potential producer side for this negotiation, Including their Chairman, CEO, CFO, R&D/Operations Heads and others.

I was, at that time, heading India Procurement for this US-based Agro Chemical company and was very much interested in bringing this business to India. This negotiation went on for few days with a lot of To and Fro communication and role play, etc. from both sides since a top person of both the companies were sitting in the meeting. Hence, everyone trying to show themselves as a strong negotiator by taking a position and were trying to convert this deal as a clear WIN for them. Face-to-face negotiations went on the whole day and eventually failed. The US team returned, and the project went to back burner. As my organisation was reluctant to go to China for this molecule, primarily because of bad experiences in the past, as they lost a lot of technologies in China.

After few months during one of my trip to Mumbai, I met CEO of that Indian company in some seminar on Chemicals in Mumbai, and we discussed what went wrong during that time and decided to meet again the next week in my office. We met and talked for quite some time and during the informal discussion, I realised that this deal is important for an Indian company to get a global footprint, as well as alliance with US MNC, is important for them for their stock valuation, which may give significant positive impact to their share prices. Knowing the fact that it is equally important for us as well to stay competitive, as cost structure in India is quite low. Beyond being the India Sourcing head, I wanted to bring this project to India before it goes to any other country. We did some trade-offs in negotiation process largely in the best interest of our respective organisations and decided to make a roadmap on how to make it happen. Accordingly, we discussed things in order to our Need Hierarchy, i.e., major things first and finally agreed to all those aspects because of which we failed in the first negotiation including price level vs. target price. Thereafter, we both communicated to our respective companies what we agreed on the following grounds and asked for their endorsement, which we got easily as it was exceeding expectations of both the companies as well as meeting their hidden needs and finally both companies were very happy.

Indian company share prices shot up in the market by 60%, when the tie-up news went to the media whereas US MNC exceeded all the benchmark and targets they set for negotiation for this deal. A few people got promoted and recognised for their business acumen ship. In short, every negotiation is dynamic. At times we need to do role play or decide whether it should be 1/1 or between the larger team, personal repo plays an important role. Also, important here is to understand key influential factor or hidden needs, it's like an ICEBERG. Therefore, as a procurement, we should explore what could be the Driving Factor or Game Changer to achieve success in negotiation.

Some of such techniques include:

The Pre-Emptive Strike – Sensing an imminent price discussion. A seller may say something like, "We don't play games with our pricing. We give a price and, if you like it, we'll do business. If not, we'll wish you luck with someone else."

The Self-Proclaimed Good Deal – The "Power of Suggestion" is a real phenomenon in psychology. By telling you that you are getting a good deal, the seller hopes that you agree and don't challenge the pricing. Of course, just because a seller says a price is a good deal doesn't make it one.

The Final Detail – Sellers often refrain from talking price until they have covered all the details and benefits of their offering. They hope that you or your internal customers will be "sold" on doing business with them before considering the price.

The Red-Tape-Wrapped Price – Salespeople are rarely the final decision makers for price improvements. Usually, prices are decided upon by their management. When salespeople sense that you're rushed, they'll make getting price approvals to seem like a time-consuming, bureaucratic task.

During a negotiation, the supplier will ask you questions. For example, when trying to decide by how much to reduce his price, the supplier may ask you: "When did you want delivery to take place?" It would be easy to say, "In two weeks" and leave it at that. I'd respond by saying, "We originally wanted delivery in two weeks, but does our delivery timeframe make a difference in our pricing?" If the answer is "Yes," I'd follow-up with, "What delivery date will qualify us for the best deal?"

Listen to Suppliers' Dialogue: Never endorse "spying" – listening to or recording someone's conversations against his wishes. But I feel that cell phone calls made in your presence while negotiating without a privacy request are fair game for listening.

In such conversations, you can pick up clues from the factors that the boss is considering revising his offer. For example, let's say you hear the salesperson say, "No. They'll be paying 100% upon completion instead of a down payment." From this, you can tell that payment terms affect your pricing. So, you can say "I overheard you mention payment terms. How might altering our payment terms help us get a lower price?"

This tactic often uncovers otherwise hidden savings opportunities. When negotiating, suppliers may mention how they are different than their competitors and, therefore, better for us. They often base their difference on common problems that buyers encounter with other suppliers. In sales, that technique is called differentiation. We can use a variation of this technique when negotiating with suppliers. It is also called "Reverse Differentiation".

In using the reverse differentiation negotiation technique, we identify common problems that suppliers have with other customers. Then, when negotiations are at an impasse, appeal to their emotions by showing how working with our company isn't as painful as working with some of their other customers.

The key is that our point(s) of differentiation must be true. If our organisation has one or more of the following characteristics, we will be different than most of our suppliers' customers and can use these characteristics to "sell" our supplier on the benefit of giving a better deal in order to earn business.

Quick Decision-Making. Some organisations take weeks or months to make a purchase decision, have it approved and place an order. If the company moves on its decisions quickly, stating how we can help our suppliers have shorter "sales cycles" can persuade them to offer better deals when negotiating.

Prompt Paying. Some organisations take 45, 60 or more days to pay their invoices, which is costly to their suppliers. If our organisation pays promptly, we can use that fact to our advantage when negotiating.

Evangelistic. Supplier marketing is more effective when their materials contain customer testimonials. But many companies do not permit their staff to offer testimonials. If our organisation willingly provides testimonials, we can demonstrate that our value as a customer is higher and that the supplier giving you a better deal would be well worth it.

Low Maintenance. Some customers whittle away at supplier profits by demanding more attention than other customers in the form of requesting special procedures, customisations or other unique attention. If we aren't a high maintenance customer, we can mention this when negotiating to give the supplier assurance that a thinner profit margin won't be eaten away by forcing the supplier to respond to special requests.

What to Do to Make a Slow Negotiation Go Faster?

Time can be your biggest lever in unclogging a negotiation that has bogged down, especially when the party you're negotiating with has more at stake in having things resolved quickly. "If time is on your side, you're at an advantage." When you negotiate a contract with a customer for a series of warranty issues, you should nudge the process along by reminding the customer of the need to keep things moving so the issues can be resolved.

If your counterpart wants to sign off on something quickly but is demanding more than you feel comfortable giving, recommending that the discussion be kicked upstairs to a manager or vice president can help resolve the impasse. When pressures are intense, a few people want to take the time to bring new people up to speed.

Keep in mind that your effectiveness in negotiating depends a great deal on your credibility. Reminding your counterparts of a time issue may speed things up, but only if they believe you are sincere and not setting false deadlines just to bring out a few more concessions. So, while time can be your ally, truthfulness is always your friend.

Do You Find Price Increase Hard to Negotiate?

How do you negotiate price increases? When should you begin preparing to negotiate price increases? The answer to that second question is, you should have begun preparing to negotiate a price increase when you originally obtained the current price.

Before I elaborate, consider a typical supplier justification like, "We must raise your price by 28% because aluminium costs went up 28% last year."

That point is tough to argue if you're not prepared. So, when first obtaining quotes for high-annual-spend products or services, ask for suppliers' cost breakdowns.

A cost breakdown will indicate the percentage of the total cost that is comprised by each major material, other materials, labour, overhead and profit. For example, a cost breakdown may look like this:

- Wood = 20%

- Aluminium = 7%

- Other materials = 3%

- Labour = 45%

- Overhead = 13%

- Profit = 12%

If a supplier proposes a price increase and tries to use the above type of justification, you can say something like: "Aluminium increased by 28%, but aluminium only comprises 7% of your price. Considering nothing else, your price should only go up by 2%."

Should you stop negotiating there?

You can argue that productivity gains should have reduced labour costs enough to offset the small material price increase. Maybe you can even convince the supplier that productivity gains more than offset the materials price hike and that your price should go down.

The key is to get that cost breakdown when you first obtain pricing from a supplier. Suppliers can be hesitant to share such information at times but are usually more willing when they are competing for your business.

A skilled negotiator knows the benefits of preparation. Here are four ways a skilled negotiator prepares:

1. ***The Skilled Negotiator Knows His Counterpart.*** Purchasing professionals often fail in negotiations due to being caught off-guard by the experience and/or aggressiveness of the supplier's negotiator. So always learn about your counterparts before you begin negotiations. Insist on a phone conversation prior to the negotiation. You can tell the supplier that the purpose of the call is to shore up logistical details like time, location and length of your meeting. And do shore them up. But also find out more about your counterpart through "small talk". How long has he been selling the product or service? Is he an aggressive personality? Then, adjust your tactics for that type of counterpart.

2. ***The Skilled Negotiator Uses Deep Logic.*** Logic can be a powerful negotiating tool. But a skilled supplier negotiator will anticipate your logic and shoot it down. For example, let's say you were buying used aircraft parts. You may say to the supplier who bid $3,000 for a part saying, "I always see these parts selling for $2,000, so your price isn't fair." That might be a good logic, but the supplier may say, "But those parts are in 'repaired' condition rather than 'overhauled' condition and don't have the same warranty." If you didn't go deep with your logic and consider all possible supplier responses, you are likely to have no more ammunition for persuasion.

3. ***The Skilled Negotiator Controls the Meeting.*** Salespeople are taught to control meetings. In a negotiation, this disarms you and prevents you from reaching your negotiation goals. Don't let the supplier control the meeting. Either present an agenda or prepare a set of probing questions to lead the conversation. And when you've

achieved the results you're looking for, give signals that the meeting is over (e.g., stand up, say "Thank you for your time in meeting with me today," etc.)

4. ***The Skilled Negotiator Knows What the Supplier Will Ask and How to Answer.*** At the outset of negotiations, suppliers want to learn if they can earn your business with their current proposals or if they must improve them. Be prepared for their questions and know how you're going to answer them. What the supplier wants to learn is:

   - Are you the decision maker?

   - Do you have the budget to pay the current price?

   - How quickly do you need to make a decision?

   - Are there other suppliers aware that you're quoting?

While negotiating, you want to end up with the Ultimate Contract – an agreement representing the best offer in the market for all terms: price, warranty, delivery, payment terms, etc. What often happens is that one supplier offers the best price, a different one offers the best warranty, a third offers the best delivery, etc.

You should never approach a negotiation feeling that you must sacrifice a good deal on one term for a good deal on another. Use this process for negotiating the Ultimate Contract.

1. ***Summarise RFP Responses.*** Use your e-Sourcing software or create a spreadsheet to list each major term and each supplier's offer for that term.

2. ***Identify the Best Deal for Each Term.***

3. ***Create the Ultimate Contract on Paper.*** Create a single sheet with the best offer for each term as if a single supplier offered all those terms. That is what the Ultimate Contract will look like. This sheet will represent your goals for negotiating.

4. ***Decide What You Can Sacrifice If You Must.*** Sometimes it's just not possible to get the Ultimate Contract. So, you must prioritise the terms you need and the one you simply want. For example, can you accept a shorter warranty if you get an even better price? Can you forgo that liquidated damages clause as long as you get

the desired payment terms? I'm not suggesting approaching the negotiation in a softer way – always strive for the best – but simply to know what the most important terms are.

5. ***Negotiate Ethically but Confidently.*** Because at least one supplier has offered you the deal you are asking for, related to each individual term. You can feel confident that the market can bear your demands. So, approach the most attractive bidder with your demands. Just be careful not to cross the ethical line. Do not disclose that you are requesting certain terms because a specific competing supplier has proposed it. But if a supplier suggests that your demand for a certain term is unreasonable, let him know that you have a proof that it is reasonable.

How Do You Negotiate When Inflation Is Rampant?

Inflation provides an easy excuse for even the smallest suppliers to try and raise prices. Suppliers often see inflation as a golden opportunity to increase margins and make up for the preceding lean years where they have been forced to give you price concessions.

So, what do you do when markets are rising and negotiating power swings in favour of the sellers? Do you succumb to the widely held belief that pricing control and predictability go out the window in inflationary times?

No way! There are several negotiating tactics that you can use to achieve cost containment that is better-than-market performance, even when prices for everything seem to be going up.

Improve Communications – Providing your supplier with better and more robust information can reduce perceived risks and lower your costs. Share real-time sales data, inventory levels and production output figures with your supplier to increase trust and eliminate the need for your suppliers to hedge.

Commit to a Longer-Term Deal – In inflationary times suppliers have a greater ability to choose their customers. You can become more important to your supplier by committing to a long-term agreement. You can also deepen the relationship through joint process improvements targeted at reducing waste or improving yields.

Ensure Future Protection – While agreeing to a long-term deal can make you more valuable to your suppliers and earn you the associated benefits, agreeing to a long-term fixed-price deal just as the market peaks can be one of the worst mistakes you can make.

Commodity prices are often cyclical. Negotiate the right type of price adjustment clause into your contracts so that today's price hike doesn't become permanent. Contract clauses should be structured to reflect the specific type of direct material being sourced and should allow for bi-directional price adjustments so that you can receive cost concessions when the market slows.

There are always things you can do to improve your situation.

The opportunity to work with your suppliers on reducing the soft costs of doing business together exists irrespective of market conditions. Getting your suppliers to agree to increase the services they offer your company or decreasing transactional costs through order automation, consolidated shipments and similar process improvements are great ways for you to add value to your company, even when inflation is running wild.

What You Need to Reach a Satisfactory Agreement

Managers frequently face the challenge of how to resolve varying points of view, so projects get accomplished, goals are met, operations run smoothly and people work well together.

Negotiation should not be substituted for decisive leadership, which involves making difficult decisions, providing clear direction and getting things done. At times when the balance of power is even, and the issues are sticky, negotiation can be an effective way to bridge the differences and achieve desired results. This is especially true when dealing with customers, job candidates and other external parties.

Successful negotiation requires a combination of skills, but it is worth learning to identify and use them. "If you do it right, you end up with a better solution for your organisation and the customer."

Know Before You Go

The road to success begins long before you open the conversation.

"Negotiating is all about anticipating. You must plan for each obstacle along the way and be prepared with the right information to overcome it. In other words, you should experience the whole negotiation before it even starts."

Whether the subject of the negotiation is scheduling a vacation, allocating work among your team or your colleagues or satisfying customers, you have to anticipate the questions your counterpart will ask and the fallback positions they will accept. Before you begin to think about your counterpart's concerns, you have to know all the details of your own position.

"The key to success is understanding your side first and your counterpart's side second."

What's involved in doing your homework? Basic preparation.

For example, you are contemplating bringing on a new hire and you anticipate a discussion about salary. You will want to arm yourself with all the relevant information: average salary within the industry, average salary within the department, average income in the geographic area, how these figures are affected by the candidate's experience and the degree to which you can negotiate. In addition, you will want to gather convincing examples to back up your position.

"It's being prepared with the information that supports your argument."

You will also want to know which data is non-negotiable, such as corporate policy on sick days or vacation days. By presenting this information upfront, you not only avoid unpleasant surprises, but you also strengthen your credibility.

"It's important to be honest and trustworthy, and for your counterpart to believe that everything you say is the truth. This is key in a successful negotiation."

Just as important as understanding your own position is understanding your counterpart's issues, concerns and negotiating style. This involves a different type of preparation, which can range from learning about a customer's past demands to having the kind of casual conversations that establish familiarity.

Start by building a relationship way before the negotiation. I do that with my customer base, employees and people in the office. I'm on the phone a lot talking to people all day long, so I get tidbits of information that I piece together.

Having a relationship provides a context in which you can more easily evaluate your counterpart's concerns. We were trying to close a deal, and there was a personality conflict between the salesperson and the customer. The customer was absolutely digging his heels in. By asking questions and evaluating the situation, we were able to get to the root of the problem – the customer was upset because our salesperson was smoking. As soon as we resolved the issue, we closed the deal.

Keep It Simple

It shows how easily negotiations can blow up into stormy confrontations and how with the right techniques they can subside just as quickly. That is why it is important to remember that despite all the complexity and gamesmanship that surrounds the topic, a negotiation is nothing more than a discussion. There are usually two points of view, both of which need to be heard. Then at some point, an agreement has to be reached.

Begin the discussion by establishing common ground. Determine why you are having the discussion and what you want to accomplish. Are you trying to convince your supervisor to let you hire another worker? Does a colleague want you to complete your part of a project earlier than expected? Are you structuring a deal with a customer or supplier?

"The starting point is making sure you have a clear understanding."

Always be thorough. Cover every aspect that will affect the endpoint you are both trying to reach. For example, if you are negotiating a contract with a customer or a supplier, "You don't just talk price, you talk terms and conditions, because if you don't have a clear understanding of the terms and conditions, there may be an unplanned cost."

Also, don't ignore areas that could be left open to interpretation.

Someone might be thinking there's an unlimited warranty when that's not the case. That misunderstanding could add cost and affect goodwill later on. To avoid that situation and to limit your liability, it is suggested to be as specific as possible.

The same advice holds true for any other discussion with your manager, co-workers or team members. You want an understanding of exactly what the other person means, otherwise, you may think you are negotiating one thing, and it turns out to be something else.

Don't Be Afraid to Ask Questions

What often derails a negotiation is when people have prepared an argument and stick to their position, no matter what.

"The ability to figure out what's really giving the other person anxiety is when you can come up with the compromise and probably win."

Achieving that understanding and using it to help the negotiation progress is a function of two key skills: listening and asking questions.

Coming to an agreement often involves a lot of emotion. Listening can defuse it.

"I wanted to let the other person know that I wanted to understand his position, so I asked questions to show that I understood his perspective," he says. "Even then, it was all about listening, because I had to learn about what the person had gone through before. Only after I got him laughing about some of the things that had happened did I present the organisation's perspective."

Asking questions cannot only reveal deeper issues, but it can help move the discussion forward. By asking questions, you reconfirm the progress in the negotiation process.

I might say, "Just so that I'm clear, let me play back what I think I heard. If it's not clear, then we circle back and start over until we get the gate closed."

The Art of the Compromise

At heart, negotiation is like chess or any other game of strategy.

"You are always thinking three moves ahead and figuring out what the alternatives are going to be when you see the other person's moves."

Always puts together a business case that shows his range of giving and takes. With a customer who is concerned about costs, you might show how different prices will affect the return on their investment. When facing a tight deadline, you might point out that while you can meet the date with the information you have on hand, if you had more time, you would have access to much richer material and would deliver a more valuable product.

"You need to know your limits and the results of your decision."

When in doubt, ask your counterpart for *his* or *her* input.

When you've tried everything, you could, but it's just not the answer you were seeking, you should ask the other person for their suggestions, which also shows your willingness to try it a different way.

Eyes on the Price

So much has been written about the concept of "Win/Win" that novice negotiators assume such results are almost given. Experienced managers, however, agree that while a Win/Win solution is always desirable on a day-to-day basis, it's just not realistic to assume that every negotiation will have that result.

Sometimes you must check your ego at the door and focus on the true problem, which is how to do the best job for the customer and the organisation. There have been lots of times when I made an assumption going into the discussion and once, I understood the facts better, I changed my mind. If you're personally invested in "It's got to be my way", you can lose sight of "what's right".

What's clear is that negotiation is not a one-time event. It is the evolution of a relationship. Even if you don't "Win" one negotiation, how you handle that "Loss" could ensure the successful outcome of future negotiations.

There are a few common negotiation mistakes we make, and it's very important to ensure that we make deliberate attempts to overcome these mistake and barriers in our day-to-day business;

- Neglecting the other side's problem
- Letting price bulldoze other interest
- Letting position drive out interest
- Searching too hard for common ground
- Failing to correct for skewed vision
- Neglecting best alternative to a negotiated agreement

The major barriers to successful negotiations could be:

- We fail to separate the people from the problem
- Entrenched positions

- Failure to review performance

- Seeing things as black and white

- Not knowing what you want and are willing to give up

- Talking yourself out of it before you start

- Using shady tactics, trickery or manipulation

- Lack of supplier knowledge

International Negotiations–Understand What Works Where

As there are different culture and styles across the global, negotiation process is no exemption. People follow a different process for negotiation in various part of the world. If we take the example of the world's two biggest business economies to understand what is different when we talk about the negotiation process in US viz-a-viz China,

Something Common in the US	Something Common in China
Quick Meetings	Long Courting Process
Informal	Formal
Make cold calls	Draw on Intermediaries
Full Authority	Limited Authority
Direct	Indirect
Proposal First	Explanations first
Aggressive	Questioning
Impatient	Enduring
Forging a "Good Deal"	Forging a long-term relationship

Country-Wise Strategic and Tactical Guidelines

ARGENTINA

Argentina is one of the most affluent countries in South America and on the continent, it's second only to Brazil in GDP per person. Food processing, with an emphasis on beef and grain exports, is a major industry and there are also considerable manufacturing and tourism sectors. It's the most heavily European influenced of the Latin American countries, and these influences carry over into the negotiation styles.

Although Argentina has a history of inflation and excessive state intervention, its new government is committed to a suitable currency and foreign investment. The government has also put a new emphasis on attracting technology firms to this formerly agricultural country. Spanish is the primary language (for business and contracts), although English and German are widely spoken.

NEGOTIATION CULTURE

The Argentinians see few rivals for themselves in the realm of sophistication. This is conceivable from a culture if not commercial viewpoint. They're hard bargainers, willing to dig in their heels if they feel they're not being treated as equals. Concessions are granted in small increments, even when the Argentinians are bargaining from a weak position. Every point of the agenda will be fought hard. Come prepared for a long haul. Although there has been a great deal of privatisation in recent years, there are still pockets of entrenched bureaucrats. Proceeding with discretion and keeping a low profile will do much to limit interference. Argentina is a land of connections and a wide range of inside tracks, ranging from family to the industrial sector to politics. The overlaps between one's social and business life are substantial, and a recommendation (or condemnation) carries a lot of weight here. Make contract as far up the organisational chain as possible. Hierarchical is the predominant management structure, with very little decision-making power filtering down the chain of command. Starting in the middle will only prolong the process and extend your negotiating budget.

While written contracts will be detailed, Argentinians prefer to do business with friends. Visitors who exhibit a cultured interest in Argentina and its people will have an advantage at the negotiating table. The meeting will be very formal and pecking orders are observed for seating and introductions. While meals may be a part of the protocol, business is rarely discussed during meal times.

Be punctual, but don't become upset if your Argentinian counterpart exhibits a more casual approach to time. Visitors, whether buying or selling, are held to a higher standard. When discussing your own company or business culture, take care to not make unfavourable comparisons with Argentina. And under no circumstances should Argentina be lumped in with the rest of South America. Politics is a very sensitive issue, and the country's militaristic past is a topic to be avoided at all costs. In all cases, the fewer political references the better.

AUSTRALIA

Australia is a western-style democracy. Over the years, its various governmental administrations have in their turn embraced Asia and shunned it. Because it's 95 percent Caucasian, it's often seen by its Asian neighbours as a western outpost. As a well-educated nation, it has one of the highest per capita GDPs on the Asian Pacific Rim. Labour unions are quite strong. Mining and manufacturing are key industries, although agriculture is a mainstay. Australians are very active as investors throughout Asia, and they export a great deal of food to neighbouring ASEAN countries and China. Foreign investment is welcome.

Australia is a technological leader in the Pacific, especially in telecommunications.

NEGOTIATION CULTURE

Australians shun formality and are recognised as some of the friendliest business people on the planet. This belies a tough bargaining ability. They're masters of the social strategy. Be direct while negotiating, as the Australians are keen to spot deception and they have no qualms about walking away from the table if they feel you're holding back information. Since formalities are minimal, negotiations move at a quick pace. Show up on time and come prepared. Australian managers tend to be more hands-on than most, so technical details will be welcomed and understood.

Even when bargaining from a very strong position, avoid the appearance of taking control. Attempting to "lay down the law" will only create resistance.

The Australians will understand the secondary nature of their own position, and reminders of it could foul the deal. The Australians don't mind putting on a little pressure when they're buying or investing. The limitations of your proposal and the availability of competitors will be cited on a regular basis. Waiting for the price to drop is an Australian pastime. Keep your offers realistic but leave yourself some wriggle-room. The Australian will haggle, but only to a small degree. Making a hyperbolic offer at the start will not be perceived as the opening of bargaining but as an indication of your lack of realistic goals.

Contracts will be written, detailed and enforceable. All parties are expected to adhere to the letter of the contract, as the Australians have well-developed commercial law. Handshakes are an amenity; signatures mean business. Because of their relatively small population and remote local, the Australians have become experienced travellers and negotiators. They do detailed research on target economies and companies, with an eye towards limiting surprises at the table. Be assured that they'll know all about your company and culture before the first meeting. The Australians are a tough breed and they enjoy competition. They never shy away from confrontation and will go toe-to-toe with anyone. While their outlook on success and failure is somewhat fatalistic, they encourage long-term relationships and prefer to work with people they count as friends.

BELGIUM

The city of Brussels is considered by Belgians, as well as by many other continentals, to be the "Capital" of Europe. All major economic and political decisions regarding the European Union flow from there. Antwerp is the third largest seaport in the world, and it guarantees Belgium's leadership in international trade and transport. The country has a strong engineering component and a robust steel industry, and it's a top player in the world's diamond market. Because of its role in European shipping, transportation equipment also accounts for a of its exports. The country is a hybrid of Dutch and French cultures, though the two seldom meet in this country of ten million. This conflict seems inexplicable to outsiders but can have an enduring effect on business negotiations and social gatherings.

NEGOTIATION CULTURE

Though socially the Belgians tend towards the casual, business meetings are somewhat formal. First names are rarely used, except among friends; Mr., Mrs., and so on is customary. French or Dutch equivalents can be used as well. The Belgians are technically astute and prefer presentations with a great deal of factual information. Get right to the point and avoid trying to sell the concept. The facts will have to speak for themselves. This is the land of commercial pragmatists; outlandish schemes will receive scant consideration. Proven technologies and recognised services tend to do the best. Do detailed research into the background of the company you're contacting. Written materials should be translated into the appropriate Dutch or French. Don't under any circumstances confuse the two cultures when negotiating, as it will cause considerable conflict. Belgians are generally recognised as tough but likeable negotiators. They've established themselves in Europe by being able to understand and absorb other cultures. Belgians have little fear of economic colonisation and always bargain as equals. Don't attempt to throw your weight around or otherwise remind the Belgians that they're a tiny country in a very large world. They've heard it all before and can afford to say "No" to just about any deal. Because of this nation's limited resources and land area, thrift is highly prized. Proposals that squander assets, whether financial or personnel, will be highly criticised. Keep it conservative and watch those pennies. Set up meeting far in advance and confirm them upon arriving in-country. Be punctual, you can assume your Belgian counterparts will be.

Though the Belgians pride themselves on their multi-lingual tendencies, contracts are enforceable only in the local dialects of French or Dutch. English is widely spoken, as is German and a multitude of other European languages. Beer production is a national treasure, and business visitors can be expected to enjoy the social aspects of its consumption. The Belgians drink for taste rather than quantity, and they appreciate a fine palate. Visitors who can hold forth on brewing varietals and production methods will find themselves in good stead at the negotiating table.

BRAZIL

Brazil is the largest country in South America, with a GDP per capita that is almost twice the size of China's. Primarily an agricultural society, Brazil exports a great deal of beef, leather and textiles, but it also supports flourishing chemical and consumer goods industries. Plagued by recurrent debt and inflationary problems, the government has instituted numerous austerity measures to keep the country on track. The communications infrastructure and high-tech industries are a target area for investors and government-backed schemes alike, and they'll remain so for the next decade.

NEGOTIATION CULTURE

Titles are very important in Brazil and business conversation remains formal. Calling a counterpart by their first name may be reserved for social occasions or for business after a long-term relationship has been established. Brazilians are very proud of their language (Portuguese) and resent being confused with their Spanish-speaking neighbours. Never refer them as South Americans and never attempt to speak Spanish in place of Portuguese, although most business people will understand both. Translate business cards and presentation material into Portuguese but avoid speaking it unless you're fluent. Brazil is a very lively place and the people there are emotional. Presentations should have a bit of style, and the presenters should exhibit a good deal of excitement about the proposal. Sticking to the facts and figures will not impress counterparts, whether buying or selling. Conversely, don't be surprised when your Brazilian counterparts erupt into emotional exchanges among themselves. Such passion is not to be seen as outright dissent but as part of the histrionics that accompany much of life in Brazil. Brazilians aren't shy about bargaining, and they tend to grant concessions only towards the end of negotiations. While the pace here may be quicker than in other South American cultures, don't expect lightning fast results. If your schedule becomes known, time constraints will be used against you. If you're unfamiliar with the commercial landscape, use an intermediary to guide you to the proper contacts. There's a very large social component to doing business and meeting the right people will not only lead to fruitful negotiations but will also ensure the long-term viability of the project. For all of their formality and relationship building, Brazilians have a reputation for loading a lot of dubious factual information into the early negotiations. It's their way keeping you interested while supplying the requisite wriggle-room if things go sour. They're expecting the same from counterparts; participation is up to the individual negotiators.

Regardless of what your hosts may say, get everything in writing, with as many specifics as possible. Never assume that you'll be able to work out those irksome details later or that business can be conducted on a handshake alone. While legal protection for foreigners is good by developing market standards, there's little advantage in putting it to the test.

CHINA

China is a fast-rising world economic power, but it still suffers from substandard GDP per capita and infrastructure. Commercial law is nascent and highly favours domestic companies. However, since little commercial law exists, a system of *guanxi* (connection) function lieu of contract enforcement. Foreign companies invest in China primarily for its cheap labour and material rates and in the hope of accessing future consumers as per capita wealth increases. The government is slowly privatising, but it still has a very shaky banking sector. Foreign firms have done well in China, although exporting profits back home is considered problematic due to China's inconvertible currency. Corruption, both public and private, is rampant and virtually unavoidable.

NEGOTIATION CULTURE

The Chinese have little compunction about deceiving foreigners. Don't rely on the other side's sense of fair play. The Chinese recognise how interested foreigners are in accessing this developing market. Investors may be led into agreements that are difficult to either execute profitably or to dissolve quickly. Profit repatriation is difficult at best and should be an early part of the agenda. Don't wait until the end to find how much you can take back home. When buying, write explicit quality requirements into trade deals, as quality tends to taper off after the first delivery.

When selling, extend no credit whatsoever. If the people who are still owed money by Chinese importers got together, they could apply for nationhood. Friendship or the appearance thereof is a common ploy to secure concessions. There may be ulterior motives behind lavish dinners and invitations to meet family members. Accept these events as gestures only. Competitors are regularly pitted against each other, sometimes at the negotiating table. Letting foreigners fight over them is a long-standing tradition here, but now the Chinese set the terms.

Many times, some form of graft will be built into the agreement. Don't assume that the person who appears to be in charge is in charge. Verify in writing, important issues before moving on to other parts of the agenda. Referring to a two-day-old spoken statement will have no weight unless, of course, you said it. Never use a translator supplied by your Chinese counterparts. If you're a non-Asian and speak Mandarin fluently, don't be surprised if the Chinese pretend they can't understand you. The difficulty of their language has been a tactical tool for centuries, and they resent having the code broken. Whenever their statements appear contradictory, it will be blamed on an interpreter. Take notes and be prepared to exploit inconsistencies across the table. They'll be doing the same. Patience is greatly appreciated, while anger and impatience are considered signs of instability. Never reveal proprietary information until a deal is completed. The Chinese don't recognise "non-disclosure agreements". In all negotiations, make it clear early on that you're ready, willing, and able to say "No" to their proposals.

The Chinese make much of the "harmonious relationship", and the leader never delivers bad news. The downside is always delivered late in the negotiations by a second-in-command, often at the banquet that follows a contract signing. Damage can be limited (not eliminated) by early statements to the effect that the contract will be subject to re-negotiation if any unforeseen circumstances change the deal's outcome.

EGYPT

Compared to the other Arabic oil-producing nations, Egypt has a surprisingly low GDP per capita, US$5,500. Its other big industries, textiles and food processing are keeping the country afloat, and the government has recently held off an Islamic fundamentalist movement that threatened another big money maker–tourism. The country has been a crossroad for international trade for millennia. But, dealing with foreign investors is quite another matter, one that's very much bound up with the nation's colonial past. While not sporting any high growth economic figures, Egypt still attracts a great deal of both trade and investment interest.

NEGOTIATION CULTURE

Haggling is a national pastime in Egypt. Foreign traders had better come prepared for some schooling how to bargain. Nothing has a set price from day-to-day; a sharp haggler is greatly respected. In fact, not attempting to bargain aggressively will be taken as an insult. The legacy of Egypt's British colonial past is a bureaucracy not unlike India's. Lots of delays, official reviews and rubber stamping of documents will be part of any medium to large project. Use Egyptian contact to help smoothen the way or circumvent officialdom, but in no way should the bureaucracy be confronted or criticised. Once it closes ranks against your project, Moses himself couldn't find an escape route.

While you may admire punctuality, your Egyptian counterpart will most likely have a more relaxed approach to time. Lateness and postponements aren't unusual, and they fit in quite well with the local tradition that whatever happens was meant to be. Arrange appointments in advance of travel and confirm them regularly upon arrival in-country. By North African and Middle Eastern standards, the Egyptians are quite familiar with modern business practices, although they don't always apply them to local enterprises. Their own system is somewhat formal and hierarchical. Rarely are first names used among business acquaintances, and titles abound.

Foreign firms are warned to have a solid agenda agreed upon before arriving at the negotiating table. All but the largest Egyptian firms will take their cues from what they perceive to be more successful foreign companies. Much of the Egyptian strategy is reactive, even when in a buying position. Contracts, regardless of detail, are considered guidelines for a business relationship rather than full delineations. The content of the document may be re-negotiated, revised and appended many times throughout the length of that relationship. Constant contact must be maintained with Egyptian partners, and their commercial "temperature" should be constantly monitored. In addition, government interference or outright interdiction in the negotiation or contract process may occur for large projects or those involving internationally recognised brands. Don't enter into business negotiations until a fairly long period of relationship building has taken place. Having the right connections and doing business with the right people can (in and of itself) prevent a great deal of delay and expense. Intermediaries may be used early in the process, but long-term and close contact must take place on a personal basis.

FRANCE

The French view their business culture as singular, much in the same way that they view themselves as special European people. Business is a means to an end, not an end in itself. The French don't often define themselves in terms of what they do for a living. Business is rarely conducted with a sense of urgency, and personal relationships are important. The French are generally considered to be a highly educated and sophisticated people. Well-read and cosmopolitan counterparts are appreciated. The economy tends towards socialism, and many of the largest industries are partially state-owned. Social issues are often a major concern when business is discussed.

NEGOTIATION CULTURE

Unless you speak fluent French, use an interpreter. Proper use of the language is a sensitive cultural issue. Even when selling, the French consider rushing to be vulgar. The more you rush, the more they'll show the process down. Don't try to out-French with the French. Even if your counterpart lacks basic social graces, allow him his sense of superior sophistication. Exploit it if you choose. Corruption is tolerated rather than promulgated. If the graft is part of your programme, be discreet. Be well prepared for every negotiation session. "Getting back" to someone will not do.

The French will discuss every point at length and will have a position on every topic. Avoid direct confrontation unless you're in a strong buying position. The French love debate but not intense criticism. It will be taken as a personal attack. All French men and women fancy themselves to be philosophers. The early meeting may have a large component of the non-business discussion. Your options will be solicited and evaluated. Not having an opinion in France is worse than having a badly formulated one. Even if you're in an extremely strong buying position, resist the urge to cut to the chase.

The French are notorious for coming to the table with a single strategy and its accompanying tactics. If they've made a completely inappropriate choice, major postponements may result while they reformulate their position. Possibly exploitable, always annoying. This is a very social culture, and negotiators may find themselves at two-hour lunches. Drinking wine will be part of the process. Keep your wits about you. You may be wearing a US$2,000 suit, but it will mean nothing if you pick up the wrong silverware or shy away from the oysters. Contracts may be long and involved, even when you are buying. All contracts must be completely in French, and commonly used foreign words cannot be substituted. This is the land where bureaucracy was invented. Red tape can be a big part of negotiating, especially when dealing with government-connected companies. Business abounds with regulations. No matter how intense negotiations become, avoid raising your voice. Doing so is a sign of poor breeding.

GERMANY

Germany's very efficient business sector is devoted to order. It has the most expensive labour force in the world but maintains a respectable productivity level. The workforce is highly unionised, and union members sit on the controlling boards of companies. As many of the country's firms have moved their manufacturing overseas to remain competitive, unemployment has skyrocketed. Consequently, the German government (as well as regional officials) has actively sought out foreign investment. The Germans are the powerhouse of Europe and the keystone of the European Union. While the absorption of East Germany has continued to be a drag on the overall economy, no one foresees the Germans relinquishing their leadership role anytime soon.

NEGOTIATION CULTURE

Codes and regulations dominate the business environment, so be prepared to comply with a rule for virtually every aspect of your project. Contracts in Germany are even more detailed than in the litigious United States. Contracts once signed, are strictly adhered to by all parties. When selling, make straightforward presentations. The Germans will be interested in the reasons for purchasing goods or services, rather than the image that surrounds the purchase. Maintain formality and observe all hierarchies.

While the Germans aren't averse to the occasional raucous social event, business-related events should be designed to maintain the dignity of all parties. Be punctual. This goes for the meeting, delivery dates, payments and social gatherings. If you're allotted thirty minutes for a presentation, don't exceed it. All meetings are planned with a start and a finish time. Germany is a land of precision; a lax attitude towards time is seen as an indication of general slovenliness.

German is a difficult and precise language. Translated material must be perfect or the content will be disregarded. The Germans appreciate anyone who has taken the time and trouble to master their language, but they tend to denigrate those who've taken half measures. You must be thoroughly prepared to answer any and all questions regarding your proposals.

The Germans have little regard for people who have to get back to them. The Germans tend towards a just-the-facts approach when conducting business. Introductions and preliminaries are brief, so be prepared to get right to the point. If your technique relies on charm and guile, this country will be difficult for you to tackle. The Germans frown on workaholics (the average vacation is four weeks).

Business is seen as a means to enjoy the finer things of life, not an all-consuming passion. Personal lives are kept separate from business. Avoid personal inquiries of the volunteering of information about your home life. Don't criticise your competitors or those of your German counterparts. Each company is judged on its own merits, without comparison to others. The Germans love to delve into the details of the proposal and shun "big picture" type presentations. They tend towards the hierarchical and view dissension in the ranks as an indication of poor preparation. Statistics and charts are a mainstay of the German negotiator and are often used to hammer home their superior technical knowledge.

INDIA

India has made great strides since departing from socialist economics in the early 1980s. Its currency, the rupee is convertible, and the heavily bureaucratic control of business by the "licence-quota-subsidy raj" has been greatly diminished, though not eliminated. All industries (except for rail transport) are open to varying levels of foreign investment. The stock market has a functioning oversight board and is generally more sophisticated than that of rival China. While it has had trouble with a somewhat xenophobic nationalist movement in recent years, India is generally considered to be one of the more welcoming of Asia's emerging markets. Predicted to have the largest population of any nation on the planet in the next decade, India is in need in virtually every economic sector.

NEGOTIATION CULTURE

India is an ancient culture with splashes of modernity throughout its business sector. Some companies run on a strict consensus basis, while others maintain more horizontal styles. Research each company thoroughly (at the highest contact level available) and find out the nature of their management style. India has numerous dialects, and contracts must be written in the vernacular or English. English is widely spoken and will most likely be the language of negotiation.

Regardless of the efficiency of counterparts, bureaucracy at local and central government levels remains lethargic. Expect delays and kick your patience into high gear. Avoid the use of your left hand for passing items– especially food. The left hand is considered unclean as it's used by Indians for personal hygiene. Be aware that much traditional Indian food is eaten with one's fingers, rather than with utensil.

Although it has numerous female political leaders, a small number of women are at the top management level in business and do alone important negotiations. Internal religious feuds and India's colonial past are very sensitive subjects. Avoid discussing them unless they have a direct bearing on the outcome of the project. Many large companies are family owned and operated, with upper management being the purview of siblings. Don't assume that good relationships with a single-family member constitute access to the source of commercial power. Internal rivalries can be bitter, personal and long-standing. Make associations carefully and only after dutiful research. Choices, good and bad, can follow you throughout this large but closely-knit country.

Indians are warm and welcoming people enjoy harmonious relationships. Confrontation isn't considered a responsible form of discussion. Like many Asian cultures, directly saying "No" is thought to be rude. Indians would sooner postpone a meeting or send along subordinates until the message gets across. India is very diverse and regionalised. Don't assume that strategies that were successful in one city will necessarily work in another.

INDONESIA

Indonesia is the largest Islamic nation in the world. Its more than 13,000 islands and its 200+ million people are ruled by a strong central authority. Much of the commercial class is of Chinese ancestry and subject to considerable resentment by the native population. The government is a mandatory player in any large project. Key industries are petroleum production, tourism, textiles and mining.

Indonesia is one of the rising stars of Asia and has considerable resources, both natural and human. Primarily agrarian until just 30 years ago, the country has made enormous strides towards industrialisation and accomplishment for which it's held up as an example to other developing Asian nations.

NEGOTIATION CULTURE

Indonesians negotiate virtually every aspect of their daily lives, from taxi rides to groceries. So, you can expect considerable haggling over even the smallest point. Part of this is a cultural norm, but much of it is an attempt to wear down the opposition. This is an Islamic culture hardly fundamentalist. Only in rare circumstances will foreign businesses have to make religious-based concessions beyond basic dietary and prayer requirements. The Indonesians are hospitable by nature but are not beyond putting a foreigner in debt for kindnesses rendered. If you attempt to reciprocate for every indulgence offered, you may find yourself giving away important concessions for very little.

Like many societies in Asia, this one is prone to hierarchy. The manager only bargains with their equals, so you may find that negotiations take place at levels that correspond directly to job titles. The greater the number of people in your party, the more complex the discussions will become. Keep it simple. While contract signing may be treated with great hoopla, don't assume that the document will be strictly adhered to by an Indonesian partner.

Contracts are simply guidelines for the harmonious relationship. Be prepared for the continual monitoring of the contract's requirements. Keep relationships healthy and avoid making the Indonesian side feel subordinate. Confrontation can cause enormous loss of face, even when counterparts are clearly in the wrong. Subtleness and finesse must rule the day. Keep in mind that Indonesian courts and arbitrators rarely produce finding in favour of foreign companies. It's either get along or go home. Bureaucracy is deep in a government rife with patronage jobs. There's no getting around it, and large projects will necessitate local input as to how to navigate through the red tape. Bribery is a standard form of getting things done, and the price goes up if the recipient must ask. These bribes may extend to the lowest level of the transaction (you often must make payoffs just to get your car through an intersection), so be prepared to grease some wheels, both big and small. Failure to negotiate a reasonable payoff will mark your company for continuous shake-downs. If your home country has severe restrictions on bribery, you may find that of–several consultation fees will be attached to your project. Budget accordingly.

IRELAND

Ireland merged from the shadow of being one of Europe's "sick men" to achieve the highest rate of GDP growth in the European Union. Many of the world's high-tech companies have flocked here to take advantage of the island's educational excellence and investment incentives. Ireland has made of the full transition from agriculture to service (now 67 percent of GDP), with many financial institutions using the workforce for back-office operations. Conflicts from British-held Northern Ireland have ceased and no longer affect the Republic. The infrastructure has been massively improved through the injection of European Union Funding, and no recipient country has matched Ireland's successful implementation. Its low corporate taxes make it a favourite of international firms and the object of scorn by more highly taxed neighbours.

NEGOTIATION CULTURE

The Irish are noted for their shrewdness and are renowned as the best horse-traders in Europe. They can drive a very hard bargain and always appear ready to walk away from the table. Ireland has been absorbing invaders for centuries and treats commercial arrivals much the same. Smiling hosts will charm even the most hardened negotiator into making concessions thought unthinkable days before. If you're charmed easily, watch out. This is a land of highly educated people; master's degrees abound at the managerial level. Highly technical presentations will be readily understood and appreciated. Any attempt to treat Irish counterparts as second rate will be met with much resistance. The error in judgement will be made evident quickly and pointedly.

The Irish language, Gaelic, is alive and well, especially in western Ireland, but English is used for most business. Government documents will be presented in both languages. The Irish are accomplished linguists, and translators for most European languages are easily obtained. These are very social people, and there is no such thing as a short conversation in Ireland. Politics, religion and many other topics are wide open for discussion. Bear in mind that the Irish are extremely well read and will know more about your country than you will.

Confrontation isn't something the Irish fear, so don't enter into one lightly, whether business-related or not. The Irish have always been clever people but have only recently had access to the capital necessary to maximise their skills. Centuries of poverty have made them very pragmatic, though not adverse to risk. Always willing to bet on a sure thing, they'll give equal consideration to a long shot if the odds are right. When making presentations, offer the upside first, but don't worry about presenting the downside.

The Irish will have no regret about calling your bluff if they think you're hiding something. They're masters of secrecy themselves and can spot it in counterparts immediately. Business and pleasure are regularly mixed, and your tongue may be loosened at the pub as part of a divisive strategy. Both guest and host are expected to buy around, though drinking is quite prevalent, getting inebriated at a social function isn't considered business-like. Enjoy yourself but stay on your toes.

ISRAEL

Starting in the mid-1980s, Israel changed its economic model from socialist to market-driven. Given its tumultuous internal political environment and its seemingly ceaseless conflict with its Arab neighbours, the country has managed to remain one of the most successful emerging markets in the Middle East. It's a leader in food production, food processing and diamond cutting, technology development. Recently both government and industry have placed a great deal of emphasis on research and development, transformation Israel into a high-tech centre. The government is very protective of domestic enterprises and issues many subsidies for start-ups. Foreign investors are rarely permitted to have controlling stakes, except in the largest of enterprises involving guarantees of technology transfers. Israel has few natural resources; imports tend to run 40% above exports.

NEGOTIATION CULTURE

Generally, Israelis see themselves as special people and are thus not bound by rules–any rules. Internally, this independence has accepted the behaviour, but it can cause conflicts in the global marketplace. Israeli negotiators will often disregard the requirements of counterparts, even when they are in a weak selling position. A clear reminder early in negotiations about who is calling the shots can prevent misunderstanding.

Israel exists in a constant state of controlled chaos. Punctuality and deadlines are demanded of foreigners but not of themselves. Israelis often attribute this approach to the country's tenuous political and geographical position and the Jewish history of rootlessness. Be prepared for the late arrival and lack of urgency to get things done. During discussions, expect to be interrupted repeatedly. While in most cultures this is considered rude behaviour, Israelis consider it a sign of interest. Israelis are very proactive negotiators and like to anticipate their opponent's moves. Unveiling your plans slowly keeps them preoccupied with guesswork and less focused on their own manoeuvres. Sudden changes in personal style also tend to spoil their plans. There's very little socialising attached to business; lengthy introductions or small talk will fall on deaf ears. Israel is a great place to do business if you enjoy a brisk pace and quick yes or no discussions. If your strategy involves long, drawn-out negotiations with numerous postponements, Israel will be a tough nut to crack.

Israeli discussions usually take the form of a straight talk called "Talking *Dugri*," which is no-frills. They'll have little compunction about calling your proposal unworkable or poorly researched. Expect no sugar coating. However, don't make the mistake of responding in kind. Israelis are very thin-skinned when it comes to foreigners and their criticisms.

Don't underestimate the role of Jewishness in doing business here. If you or members of your team are Jewish, play it to the hilt. If neither case is true, avoid discussion of religion. Don't try to disregard the fact that preference will be given to Jewish-owned businesses and their representatives. Finding and fostering the right connections in Israel will result in long-term success. Avoid discussions about Middle East politics or the founding of Israel. You may be sounded out on this topic by Israeli hosts, but it's best to avoid the bait. Keep the talk focused on business.

ITALY

Italy is often considered a problematic member of the European Union. Its volcanic domestic politics, frequent bouts with inflation, mounting public debt, high unemployment and high government spending have combined to make it almost impossible for the country to meet its European Monetary Union (EMU) goals. Despite these problems, Italy has turned in some surprising domestic market successes. Its manufacturing, auto and clothing and jewellery industries have continued to make progress, while high-tech companies are burgeoning. Italy tends to run trade surpluses (10 to 12 percent), and non-EU investors will find that acquiring an Italian partner is the most efficient way to gain market access.

NEGOTIATION CULTURE

Time is not of the essence in Italy, but punctuality is appreciated. Schedule meeting far in advance and confirm them upon arrival in-country. Only deadlines that are repeatedly insisted upon will be met. Connections are very important. This amounts to someone, or some company, giving a good reference about you to your potential counterpart. These are very personal references, much deeper than the standard letter of introduction. These connections take some time to develop as personal integrity and loyalty are at stake. Italy is highly regional.

Much research and relationship building must precede any investment move. Even with simple trading, on-the-ground research is recommended. More so than most other EU countries, Italy will require a patient outlook when approaching business deals. The Italians like to do business with not just friends but respected friends. It may take a while to gain their confidence, but the effects will be long-lasting.

The Italian manufacturing sector is known for a high-level quality and accompanying prices. While cost control is everyone's concern, any indication that you want to do something on the cheap will get a poor reception. It's all in the phrasing, and interpreters have to be well-schooled in your country's vernacular.

Italians are impressed with counterparts who share their cosmopolitan tastes. Expressing a genuine appreciation of art, music, food and fashion will keep you in good stead with the locals. Avoid the just-here-on-business approach. Count on participating in a good number of social events and lengthy meals as part of the business protocol. When it's your turn to reciprocate, seeking the aid of local hoteliers or restaurateurs are good sources.

Italians can be argumentative about virtually everything. The raising of voices and flailing hand motions are national pastimes. Don't be fooled. This is just passionate discourse, not signs of anger. It's rarely directed at counterparts but witnessing such an interaction across the table may give the wrong impression about the cohesiveness of the other team's position.

JAPAN

The "Bubble Economy" of powerhouse Japan burst long ago. Today, Japan's economy is what may be described as "Sputtering". However, the country is a long way from destitution and will hold its status as one of the top economies in the world for some time to come. Japan is slowly moving away from its more interventionist approach to more competitive market economics and is breaking down the protective walls used for decades to shield its now-powerful industries. Illicit behaviour in both government and private sectors is now regularly exposed in a culture that, just a few years ago, would have considered the process unthinkable. Most observers regard this change a sign of a more mature, less insular Japan.

NEGOTIATION CULTURE

Letters of introduction from a *Chukaishu* (well-placed go-between) are essential for doing business here. As in much of Asia, having connections in Japan is the only way to get through the front door, unless you represent an internationally recognised brand name.

Individualism is not a characteristic of Japanese negotiators; they rarely come to the table in groups smaller than three. If you're a solo act, you'll still be confronted with a team ever for simple trades. Be prepared for several informal trade barriers and restrictions on foreign investment. They'll make no apologies for their protective system. The meeting will be very formal, and the company hierarchies will be observed.

Keep in mind that the person doing the most talking for the Japanese side will most probably *not* be the person in charge of the negotiations. This permits all mistakes to be covered by speaking out of turn excuses and prevents the company from making commitments until they've seen the entire proposal.

The Japanese will not discuss points that are not part of the pre-arranged agenda. Small talk will be kept to a minimum and inquiries into personal lives will rarely be made or accepted. The meeting will be highly orchestrated; the Japanese don't like surprises. Have your presentation materials translated into Japanese and keep them detailed. Japanese negotiators are famous for their ambiguous responses to proposals. They view vagueness as a form of protection from loss of face in case things go sour.

Contracts are viewed as guidelines, and any problems are arbitrated, rather than litigated. Contracts will almost always include a *jijo henko* clause that permits complete re-negotiation if circumstances change. The system works well in Japan, but many foreigners, especially Westerners find it disconcerting. The consensus is a lifestyle in Japan and the *ringii* system of moving information up and down the chain of command can be a trust test of patience for foreign companies.

The Japanese aren't noted for being able to think on their feet, so don't expect a quick answer to any question or problem. Cost of capital is very low in Japan (often less than 1 percent), and they can afford to take a long-term approach to profit making. The Japanese have traditionally sought market share over quick returns and they expect foreigners (especially investors) to share their commitment to long relationships. Be forewarned: a quick return on investment in Japan is about ten years.

The Japanese maintain harmony at all costs, will smile the most when they're the least comfortable. At the negotiating table, a toothy grin is a sign of trouble, not acquiescence. If your proposal is unacceptable, you won't be told "No" in a direct manner. Postponements and request for further research should be interpreted as a prelude to failure.

MALAYSIA

Three decades ago, Malaysia was little more than rice paddies and rubber plantations. Today, its home to one of the world's tallest building and touting its "High-tech corridor". Essentially an Islamic (though secular) nation, it has combined the innate talent and energy of its native population with the mercantile strength of its Chinese minority to produce one of the most dynamic economies of Asia. While it still maintains strong agricultural, petroleum and rubber manufacturing components, this nation has hitched its wagon to high-tech and is positioned, both physically and economically, to become the major technical player in ASEAN. Though it suffered a serious currency devaluation in the late 1990s, the economy remains strong and attractive to investors and traders alike.

NEGOTIATION CULTURE

Malaysia has a very formal, almost ritualised, culture. In a country with a potential for numerous ethnic and religious conflicts, the preservation of face and the importance of mutual respect keep things running smoothly. The country's colonial past makes Malaysians highly suspicious of foreigners, at least initially. When in doubt, stick to the formal and overly respectful. If it's too much, your Malaysian counterpart will discreetly make it known.

As in much of Asia, the direct approach is considered rude and lacking in finesse. Pleased reactions will be downplayed, and negative ones completely covered with distractions and forced smiles. This isn't lying, per se, but a way of avoiding embarrassment for both sides. Until something is in writing, assume "Yes" means "We can continue discussions", and don't ever expect a direct refusal.

The Malaysians are a detail-oriented people; discussions may drag on as every point is exhaustively dissected. Foreigners can take such nit-picking as a sign of great interest. Short discussions mean it may be time to reschedule your flight for an earlier one. Consensus decision-making and hierarchical lines of communication will be an additional drag on negotiations. Even the most powerful head of a Malaysian company will submit to group thought. You can't rush it and you can't avoid it, so just allow time for it. Most large companies are Chinese owned, but by government edict, they must have a native Malaysian component. Most of the ethnic Chinese have taken Malaysian-style names to downplay any resentment. Foreigners should avoid any mention or exploitation of this rift, and it is a criminal offence to incite ethnic rivalry. As far as you're concerned, all counterparts are simply citizens of Malaysia. Most business discussions will involve some form of socialising, usually over a meal.

The Malaysians are rightfully proud of their cuisine and see it as a national treasure. Don't expect alcohol to be served unless the major players are non-Islamic. If your in-company stay is lengthy, you'll be expected to reciprocate with a banquet of your own. Seek local help in setting this up, as the protocol is fairly rigid. Contracts are considered secondary to the quality of the personal relationship of the principles. You must keep the relationship strong and current if you expect to succeed here. Almost all contracts contain some form of escape clause to release both parties if things go sour.

MEXICO

Mexico has been economically and politically turbulent over the last two decades. Massive currency and debt problems have been exacerbated by political assassinations and corruption at the highest levels of government. None of this, however, has seriously damaged the country's attractiveness to foreign investors. From the success and conversion of *Maquiladoras'* established just inside its borders by its northern neighbour to the growth of Japanese auto plants, Mexico continues to make a serious contribution to both NAFTA and world trade. Low labour rates and liberal (by other emerging market standards, at least) foreign investment packages further Mexico's drive for success.

NEGOTIATION CULTURE

Formality marks Mexican business at all levels. The system is very hierarchical, and communication is directed downward. Only equals address each other on familiar terms and only after a lengthy relationship. Meetings consequently take on a very stilted air and much care will be given to seating arrangements and introductions. Mexican society, both cultural and commercial, puts a great deal of emphasis on the showing of proper respect among its participants.

Slights, whether real or imagined, are potential deal breakers and Mexicans are noted for their sensitivity to foreign opinions. The smile that looks like a smirk or the fatigue that's mistaken for boredom can bring the most lucrative deal to a screeching halt. Even investors and buyers must steer clear of these sensitivities.

Relationships are extremely important here, and even short-term business will require them. The establishment of these *palancas* is part of the system of reciprocity that's at the economy's core. There's little doubt that some of this relationship building will result in graft; foreign visitors should be prepared for that contingency.

Mexico has a well-deserved reputation for bureaucratic lethargy–the result of the entrenchment of the PRI as the nation's ruling party. While the situation is improving, projects involving foreigners (especially US citizens) receive extra scrutiny. Even barring government intervention, life is slow-paced and patience will be required for even small trades. Those in a hurry are advised to look elsewhere for business opportunities.

Attitudes towards punctuality and deadlines are nowhere near the extremes of the *manana* stereotype, but Mexicans do view work and business as a means rather than an end. If punctuality or a deadline is important, that sense of urgency must be instilled early in the relationship. Mexican businessmen will readily agree to deadlines and payment schedules that (in their eyes at least) will be regarded merely as guidelines. Contracts are a matter of personal interaction in Mexico, and this attitude is reflected in the scant commercial law. Contracts are only honoured among friends; lawyers hold little sway.

Foreigners should be reminded that Mexican courts are heavily weighted towards domestic companies. Mexico is the land of *machismo,* with male dominance and gender roles placing high on the list of sensitive issues. Female foreign executives may find the going very tough, when attempting to maintain their business role, and they'll see no female counterparts across the table. Mexican men simply don't take direction from women, and they have little respect for men who do. Bucking the system will unquestionably have a deleterious effect on the deal.

PHILIPPINES

This island nation has come a long way from its status as the sick man of Asia. It has learned a great deal from the mistakes made by the other emerging markets of ASEAN and has sought growth at a manageable pace. Foreign investment laws have been revamped to grant majority ownership even in the banking industry. The country has pursued build-operate-transfer-style infrastructure investments to solve its energy and communications needs while attempting to avoid environmental degradation.

It remains an agrarian-based economy with pockets of industrialisation, and it sees itself as having a productive future in shipping and tourism due to its prime location in the Pacific Rim. The successive governments have instituted a programme to emphasise deregulations and modernisation while creating numerous Export Processing Zones and Special Development Programmes throughout its far-flung archipelago.

NEGOTIATION CULTURE

Negotiations in the Philippines are conducted in a very formal manner. Company titles are very precise and are used to maintain the hierarchy. Locals will take on a very rigid manner when dealing with foreigners in an effort to preserve what is believed as a professional appearance. The Philipinos are strong believers in forging relationships and maintaining *pakikisama* (smooth relations) at all costs. Confrontation is unthinkable and a sign of disrespect. Part of this process is the reciprocity system whereby one business connection (or political) connection leads to other, more lucrative, deals. Acceptance of a favour or reference will call for a larger one in return. Beware of Philipino bearing gifts–a simple "Thank You" will not suffice.

While the government has made a genuine effort to attract foreign investment, it still takes action to override contracts and nullify their terms. This is all part of the "Philipino First" policy that it has pursued in recent years to protect domestic industries from foreign exploitation. The policy usually comes in to play when lucrative deals are given to the more competitively price foreign firms. Having *pakikisama* with powerful people will limit your risk of being victimised by this policy.

Gossip is both a hobby and a business weapon. Competitors, both foreign and domestic, can be the subject of most ridiculous rumours imaginable. This effort to set up intrigues, thereby derailing negotiations or even contracts, is a standard procedure. Keep an eye on your *pakikisama*. Like many former colonies, the Philippines put a great deal of emphasis on maintaining one's self-respect when dealing with foreigners. If you're buying, make sure you do nothing to demean your counterparts as the revenge will be in missed deadlines and quality problems. If you're selling, learn how to bend over backwards.

Laws of all kind are very flexible in the Philippines and are applied in direct proportion to the amount of money at stake. *Lagay* (bribery) is very much a part of any business, especially big business. The money will be paid in some form or other and individual negotiators must be prepared to satisfy both their Philipino counterparts and their home country judiciary. Leave your moral baggage at home and pack extra savvy. Higher education and advanced academic degrees are held in very high esteem. If you have some academic initials to include after your name, do so and expect favourable results. When introduced to counterparts with a similar distinction, inquiries into the particulars of their education will be met with enthusiastic responses.

POLAND

Poland has moved from the sluggishness of a centrally planned economy to become the strongest commercial entity in Eastern Europe. Its growth figures are on par with the emerging markets of Asia, and the government is making a sincere effort to dismantle unproductive state-owned companies. Poland has a strong iron and steel sector, as well as formidable machine manufacturing facilities. It still imports a great deal of fuel and foodstuff, often in exchange for manufactured goods. The Poles have a dynamic culture, clear-cut ideas about their economic future, and little desire to permit a repeat of the colonisation that plagued their past.

NEGOTIATION CULTURE

While its culture and language are quite old, Poland as a nation is relatively young, and it has only recently emerged from under the shadow of the Warsaw Pact. Foreign negotiators must recognise that this is a country with a mission to make up for lost time. Don't start negotiations unless you're serious. If you're here to do research, make that clear. Poland achieved a high state of industrialisation under Soviet dominance and is now looking for technical development. Negotiators will be pressed for technological transfer, not just for investment funds and plant construction. It's best to hold this concession until all other points have been dealt with effectively.

For all of their desire to move ahead, the Polish decision-making process is awfully slow. Consensus rules the day communication is rigidly hierarchical. Much of their internal debate may take place away from the table, but it's time-consuming, nonetheless. Any postponements will most likely be the result of internal problems and not necessarily the result of a proposal's shortcomings. Foreign negotiators from longtime capitalist economies must keep in mind that properly prepared financial statement and property valuation are new concepts.

Trying to get an accurate answer about what the counterpart's company is worth or its potential for growth will be just that—a trying experience. It may seem that you're being intentionally deceived but for the most part its lack of training. If such documentation will be required, get preliminaries prior to travel so that you'll know how much work is ahead of you. Don't arrive in Poland with only a single plan for trade or investment. Develop a variety of positions that can be presented as the negotiation terrain changes.

Be punctual and be prepared. This highly educated population sees punctuality and follows through as indicators for respect. Remember, this is the former Second World, not the Third World. Socialising, especially drinking, is a part of business life. If you're not a big imbiber, keep a low profile; it's better to abstain than to make a fool of yourself. Women are rarely found at the managerial level, although the pole will not object to female counterparts.

RUSSIA

Russia is the geographically largest sovereign nation on earth, with a wealth of natural resources that alternatively suffer from exploitation and neglect. The fall of the Soviet empire left Russia with massive debts and a devalued currency that was inconvertible as late as 2006. Even now, US dollars and euros are the preferred scrip. The new Russia originally made enormous strides in its effort to throw off the burden of central planning but began to reassert state control once the inequities of privatisation became clear. Solid predictions of long-term, high single-digit growth are attracting large but prudent investments. Russia has seen the steady rise of the *mafiya* in business and the renewed scrutiny of business by the state.

NEGOTIATION CULTURE

Russians may be new to international commercial negotiations, but they're old hands at negotiating with foreign powers. They have clear agendas (although you may never see them) and no strategy or tactic is off-limits. Russia is no place of amateurs. Send your best. Arrive at the table with clear objective and hard-liner stance. Make few concessions early in the discussions, but know that your Russian counterparts will make numerous, inconsequential ones. If you play by their rules, you're doomed.

Remember that Russia needs practically everything–including you. The Russian people are extremely warm and gregarious. It's very difficult to dislike them on a personal level and this can be used against you. Parties, dinners and introductions to friends and family, while socially standard, can lull the foreigner into a belief that they'll be treated as friends at the negotiating table. Enjoy yourself but separate business from pleasure.

Much of the difficult discussion at the table will be about how the buyer pays the seller. If the Russians are selling, they'll want to be paid in advance with a hard currency which to them is just about anything but the ruble.

When they're buying, payment will be delayed, and you can guess what the preferred currency will be. Foreign investors should never, under any circumstances, agree to single block transfers of cash. Keep the transfer small and dependent on preset benchmarks.

Although it has left central planning behind, the Russian decision-making apparatus is very bureaucratic. Even the simplest deals will take a great deal of time when compared to other industrialised powers. Numerous trips will be required for medium to large ventures. Just as the geography lies between Europe and Asia, so does Russia's attitude towards contracts. It's best to get as many details written into the documents as possible, in the hope that most will be complied with during the agreement. Important points must be continually emphasised, as the Russians tend to look at the totality rather than the details of a contract.

The *mafiya* is a major player in Russian business. Foreign businesses are regularly shaken down and their personnel roughed up by its representatives. Don't expect that you'll be able to extract any kind of pressure against them at the negotiating table. They're a de-facto sub-government and must be courted accordingly. Bribery is as common as taxation. If you can't deal with this unsavoury aspect of the business, Russia isn't for you.

SAUDI ARABIA

Saudi Arabia is both the centre for Islamic culture and a wealthy petro state headed by a royal family with absolute power. Oil & petrochemical derivatives are its major industry, and its huge land mass contains only 28 million people (that include almost 6 million foreigners). Due to its wealth, much of the labour force is expatriate, imported from the poorer nations of Asia. The nation is dependent on foreign firms for most of its infrastructure and its construction. The Saudi family overseas controls all major commercial deals and has tried to move the economy into the steel and cement industries with limited success. Most private businesses are services directed at the petroleum and petrochemical industry.

NEGOTIATION CULTURE

Most of the people you'll be negotiating with will be directly associated with the royal family. While your punctuality will be taken as a sign of respect, your counterpart's lateness is a sign of power. The Saudis rarely see themselves as "needing" foreigners and do little to ingratiate themselves. The

Saudis are very wealthy but unwilling to admit to their limited business skills when it comes to dealing with non-petroleum issues.

Impassivity and stern countenances often disprove confusion. It's best to permit the Saudis their façade of shrewdness, as long as you understand its depth. Your Saudi hosts will want to size you up before getting down to business. Social banter will precede business; it should be left to the host to decide when there has been enough chit-chat. This social aspect of business requires that a lengthy courtship predate the actual deal. All business, regardless of contracts, is conducted on a person-to-person basis.

The Saudis see themselves as special people. They take a very arrogant, almost take-it-or-leave-it attitude towards foreigners. Most of the country's wealth came after World War II, and its people are still trying to overcome a past in which they were subject to foreign rule. Negotiators must understand the cultural background in order to understand their counterpart's posturing.

This is an Islamic (though not fundamentalist) state and home to the most sacred site in the religion, i.e., Mecca. Usury (the charging of interest for loans) is forbidden under Islamic law. Therefore, negotiators need to come equipped with a variety of imaginative financial packages in order to comply. Islamic banks don't charge interest but become partners in the project, thereby fully risking their investment.

Bargaining usually starts at an artificially high level when the Saudis are selling and below ground level when they're buying. Don't be put off by these sometimes ridiculous offers. It's just a traditional starting point and is used to provide the maximum room for manoeuvring. Women have absolutely no role in Saudi business. Bringing female executives to the negotiating table to wrangle with Saudi counterparts may spell disaster. If your company wishes to force the issues, do so only after performing in-depth research into the local customs regarding the appearance of women in public. Even the US military had to make concessions on this point during the Gulf War, and they were in the strongest of buying positions.

SINGAPORE

Singapore is often held up as an economic model for other Asian countries. Strict civil and criminal laws are matched by regular government intervention in commercial activity. Its meteoric rise as a commercial power in Asia was funded largely by outside groups. Singapore was considered as a

"Developing" nation during 1989 but is now seen as a hi-tech and financial service hub. It's a shipping and transportation powerhouse, and its wealthier companies are very active in investment throughout Asia.

NEGOTIATION CULTURE

Singapore is a former colony and its independence is relatively new. Singaporeans are very suspicious of outsiders, especially Westerners, and fear being busted at the negotiating table as much for the economic loss as the perceived loss of status. When you're winning, avoid any appearance of smugness. Pay close attention to details. There's scant legal protection for foreign companies here, and all loopholes will be exploited massively. Don't be shy about getting everything in writing, as your counterparts will be relying on your lack of confrontation to get their way.

The Singaporeans are practitioners of real politics and don't place much emphasis on the value of the personal relationship. While socialising may be part of the negotiation process, it's something that will be used to lull the visitor into a false sense of trust. The smiling face that treated you to dinner last night is the grimace that meets your gaze across the negotiating table today.

Singapore is a mixture of Chinese and Malay cultures, but its pace is more on par with that of Tokyo or New York. Deals often move along quickly and the tempo may quicken if they fear they're losing control of negotiations. Set a comfortable pace and stick to it, and under no circumstances should your counterparts know when you're pressed for time.

If you're selling, start high and work down. Starting off with a fair price will get you nowhere. If you're buying, assume that the price the Singaporeans are asking for is far higher than the one they expect to get in the end. Arrive at the table with many options, including the possibility that this round of negotiations may fail. Singaporeans like to drive hard bargains, and they have no mercy for counterparts who have a do-or-die negotiating mission. If they sense that you have no alternatives, expect a good hammering.

English is the language of business here, and its study is a mandatory part of education. Older businesspeople may not be as fluent as their younger subordinates. Singapore is very cosmopolitan, and interpreters for most languages are available locally. Keep in mind that misinterpretations are a standard way for Singaporeans to cover mistakes in their tactics.

Contracts are taken very seriously in Singapore and aren't meant to be broken or continuously re-negotiated. The contract may also contain language about how to maintain harmonious relations with your new Singaporean partners. Corruption is minimal, especially when compared with the rest of Southeast Asia. The country has a very high GDP per capita and bureaucrats are well paid. Foreign negotiators offering bribes are dealt harshly. This country is law-abiding by the nth degree. In fact, some expats have nicknamed Singapore "Disneyland with a death sentence".

SOUTH AFRICA

South Africa spent many years in a state of International embargo and divestment by foreign firms. The end of apartheid had seen the return of investment and the restructuring of the nation's industries, but results have been far from stellar. High crime rates coupled with the induction of inexperienced managers into top positions has made South Africa decidedly unattractive to all but the most altruistic. It's rich with mineral resources and even in its current state continues to be the most industrialised of African Nations. South Africa is also touted as having a bright future in shipping, due to its location. Many investors are waiting for the cultural and political dust to settle before making a final assessment of the nation's potential.

NEGOTIATION CULTURE

The business atmosphere is as racially charged as South Africa's politics. Many new managers and CEOs are black and view their role as establishing a new order. Some take a very self-righteous view towards investors, whites in particular. Visitors may play to this perception for advantage or downplay it together. Confronting or disparaging it will have disastrous effects. White-run companies, for their part, refuse to be held responsible for the history of apartheid and view opinionated foreigners as meddlesome. It's best to keep opinions on racial topics to yourself.

South Africans of all colours are somewhat new to the international arena and many believe that the outside world needs them more than South Africa needs foreigners. Don't be surprised if you encounter "Take-it-or-leave-it" attitude across the table even when you're buying. All negotiations should start off with a clear picture of how both sides stand to benefit from the completion of the deal. This is a nation used to self-sufficiency, and its business people tend to bristle when they feel they're being pressured.

Trying to apply artificial deadlines or ultimatums will only further delay the process.

South Africans like straight talk and low-key sales approach. When South Africans are in a strong buying position, they'll intentionally slow down the pace of talks in order to extract concessions. Socialising is a big part of the South Africa business protocol. Attending sporting events, after-hours partying, back-country tours and even hunts are used as a way to size up foreign counterparts. If they don't like you personally, you may find it difficult to make headway at the negotiating table.

This isn't a litigious society, and business people here don't like to quibble over details. Bringing an army of lawyers to the table will not be of great benefit. Contracts are composed very much along European lines, and South Africans prefer to allow the quality of the relationship to fill in any vagaries. English is the language of commerce; interpreters for many European languages are readily available. Africans business people, who speak a form of Dutch in addition to English, will often bring their own translators to meetings with foreign companies.

SOUTH KOREA

The Republic of Korea (ROK), as opposed to the Democratic People's Republic of Korea situated due north, has made enormous strides in the last few decades and is now one of Asia's powerhouses. It's striving to become more of a global player and less of an Asian one. Modelling themselves on the success of their rival Japan, the South Koreans continue to use their central government as the overseer and occasional planner of the nation's major industries. But moving quickly from an industrial base to technology has caused the South Koreans to suffer growing pains. The country has a number of Special Economic Zones set up to entice foreign investment as well as hoped-for-technology transfers, despite its outward-looking commercial policies.

NEGOTIATION CULTURE

South Korea still maintains an air of siege mentality in its approach to foreigners. These are tough people who will go toe-to-toe with anyone regardless of size. South Koreans have zero reluctance about signing agreements with which they have no intention of complying. Foreigners are meant to be fleeced and there's no shame in it. If you sign a contract in the ROK, be prepared to supervise its compliance on a regular basis.

Wearing down an opponent through constant repetition and lengthy negotiation session is a time-honoured practice. It's also a way of making sure that every detail has been covered. Mistakes by Korean managers are dealt with severely, especially if they cause the company to lose face at the hands of a foreigner. Being direct isn't considered a Confucian virtue; thus, such behaviour is incompatible with Korean business practices. Even if your counterpart is telling a bold-faced lie, diplomacy must be your response.

Koreans often make emotional pleas part of their negotiating style. They're also not beyond painting themselves as poor, humble peasants, even though they have one of the higher GDPs per capita in Asia. If you demonstrate any compassion, it will be tapped again and again with no hope of reciprocation. Koreans only respect hard-liner, strong opponents. Koreans always negotiate in teams, and they'll always attempt to be numerically superior. Exploiting dissent or contradictions in foreign counterparts is a common ploy. Unless your team is composed of highly skilled Cowboys, don't let anyone gets separated from the group. The weakest members will be culled first, so keep a special eye on novices. Much of the ROK's success is the result of sacrifice by the general population. Koreans work hard and play hard. Negotiations will be exhausting and after-hours socialising will be more a test of fortitude than a chance for relaxation. If you're not willing to *gun pei* (raise the cup) until the wee hours with counterparts, it's best to decline altogether. Women play a very subservient role in Korean business and will only appear at negotiations as secretaries or interpreters. Visiting female executives will have to make a special effort to make their status known.

SWEDEN

Sweden is the most socialised country in the world, with over half of its GDP being in the public sector. Consequently, it also has one of the highest income tax rates in the world at 60 percent. Sweden's once highly touted social welfare system, *folkhemmet*, has taken its toll on the economy as domestic firms move operations abroad. Its major industries are steel, automotive, precision machinery and wood products. This densely populated country is very dependent on foreign oil and imported foodstuffs. Foreign investors, though continually attracted to the highly skilled and educated workforce, are often deterred by what is viewed (at least by outsiders) as confiscatory taxes.

NEGOTIATION CULTURE

Make an appointment well in advance and be punctual. Rest assured that Swedes counterparts will do the same. These are precise and disciplined people, and they expect the same from those wishing to work on their turf. The high level of education makes technical acumen quite common, and the Swedes are known for their engineering skills. Presentations should be detailed with additional data at the ready. Keep proposals realistic and play up the pragmatic aspects. Exaggeration or melodramatic presentations will be met with classic Scandinavian indifference.

The Swedes are tough bargainers, but they rarely haggle. Their approach is to remain staunch in the own position while constantly demanding concessions from counterparts. They will comb through the details of your proposal looking for loopholes. Be prepared. This is a very formal, law-abiding society and bribery is completely out of the picture for negotiations. If your strategy relies on conducting business under the table, you'll find yourself alone under there, or worse, incarcerated.

Upper-level Swedish managers often delegate a good deal of–decision-making authority to middle-level management. Don't assume all power is at the top and do your research to make sure that you have decision makers sitting across the table from you. Many Swedish companies use the executive suite as a preretirement holding area. A company president may be so in name only. Contracts are very detailed; expect them to be followed to the letter. The Swedes will expect the same from counterparts. While this seems quite efficient, be reminded that only that which is in the contract will be attended to and little more.

Swedish counterparts will most likely request to see your entire proposal before coming to the negotiating table. This is not just to limit surprise but efficient use of time. It also serves to avoid potential conflicts or confrontations–both highly undesirable here. The Swedes like to combine business with pleasure but not of the raucous type. Business meals will be formal, and toasting will be moderate. Women makeup 48 percent of the workforce and hold high positions in government and business. Contrary to the image of sexual abandon, Swedish men maintain a rather traditional attitude towards women in society. Foreign female negotiators can expect to be subjected to a good deal of Old-World manners and "Ladies first" deference.

TAIWAN

Taiwan, along with Korea and Japan, was one of the early Asian miracles of the 1960s and 1970s. Its commercial sector is still a leader in the global market, and its GDP per capita is five times higher than that of its larger cousin across the straits. Taiwan is both an industrial and a technical power that has seen a sharp decline in foreign investment in the last few years, due to higher labour rates and uncertainty about its relationship with mainland China. To hedge its best, Taiwan has poured investments into China and its neighbours. The Island's business people are highly schooled in international business, and thousands of them are sent throughout the world, on a regular basis, searching for opportunities.

NEGOTIATION CULTURE

The Taiwanese are more expert at international negotiations than any other Asian group. The insular nature of their economy and their tenuous political position has kept them on the global scene for decades. They'll know more about your society, culture, business and company than you will. Similar research on your part is vital. Learn everything you can about your counterparts but be discreet when tapping for information in Taiwanese business circles–they're quite clannish.

Expatriates with on-the-ground experience are the best source. The Taiwanese are Chinese by heritage but "Special Chinese" by circumstance. Attempting to treat them like naïve mainlanders or cosmopolitan Hong Kongers will be met with disappointment. These Chinese are believers in real politics and have no delusions about their position in world economics. If they're buying, expect a tough stance; if they're selling, be prepared for a healthy dose of indulgence, but keep your hands-on your wallet. Come to the table prepared.

The Taiwanese are highly educated, more so than Hong Kongers, and they take business very seriously. Poorly prepared counterparts will be quickly shown the door, though very politely. Under no circumstances should you go on a fishing expedition under the guise of true negotiations. While the Taiwanese aren't beyond doing these themselves when overseas, they frown on turnabout. Foreigners who are caught doing it will find it difficult to get appointments anywhere on the island when they wish to negotiate in earnest. News travels fast.

The Taiwanese tend to front-load their concession process with numerous, inconsequential ones early on. Bigger concessions are saved for the end and given with great reluctance. The Taiwanese often practice the envelopment of opponents. You'll be wined and dined, hotels (perhaps offered airfare) and drivers will be provided, and you may even be taken to a remote resort for the negotiations. All of this puts you in debt to your hosts. It also makes you psychologically dependent on them for even the simplest things. Enjoy the ride but stay on your toes.

Visiting foreigners are often amazed by the amount of after-hours partying that accompanies business here. Your hosts will most likely pull out the stops during negotiations to wear you down. Remember, they can always find new team members for the following day's discussion, and none of their team will be suffering from jet lag. Take it easy and when in doubt, call it an early night.

THAILAND

Thailand has been one of the "Mini dragons" successes of Asia. It has long been a target for foreign investment, especially hotels and resorts, and tourism is the major source of foreign currency. The economy also supports sizeable textile, agricultural and aquaculture industries. Thailand's numerous banking industry scandals have caused the country to seek international bank funding. This has been brought with it a great deal of external scrutiny and the rethinking of many projects involving foreign private funding. The downturn in the high growth rate of Thailand (less than 5 percent) and continued political chaos have had a chilling effect on FDI here and in the regional emerging markets as well.

NEGOTIATION CULTURE

Personal relationships are a very important part of doing business in Thailand. Having the right connections and knowing when to use them will greatly facilitate long-term success. Start working on these relationships well in advance if attempting to schedule negotiations. Patience is highly regarded in Thailand and Buddhist countries in general. Thais believe in keeping a cool heart during negotiations and meetings. Showing open frustration or making demands to speed up the proceeding will not a show of strength but rather as a personal weakness.

Any concerns in this area should be discussed via back-channels, never in public. It's very important to maintain a united front when dealing with Thais. Although they'll be just as willing to exploit discrepancies as the next negotiator, disunity also carries the burden of disharmony, which will cause Thai counterparts to doubt the project's potential. Punctuality is very much appreciated, although its appeal declines in proportion to the distance one is from Bangkok.

Agendas should be discussed and agreed upon prior to arrival in-country, but don't be surprised if your Thai counterparts suddenly attempt to add or lessen from the original. Spontaneity is a big part of Thai business culture. Bargaining should be done in such a manner as to be obviously concerned about all sides of the deal. Any appearance of haggling just as a show of superiority will have detrimental effects. Proceed as if the relationship was more important than the profits. This is a difficult language to master, and your attempts at it will be appreciated.

While English and Chinese are widely spoken, translate all materials, including business cards into Thai for presentation. Since Thai speakers aren't all that common outside of the country, this procedure may have to be done upon arrival in Thailand. Thailand is a functioning monarchy, and the people are devoted to the royal family. Criticising or insulting the king or his family can result in criminal charges. Very few (if any) Thais will take kindly to a foreigner making even the smallest critique of the monarchy. Graft and corruption are part and parcel of the commercial sector. Requests for "Tea money" and the inclusion in projects of the inflated consulting fee may be part of the price of doing business in Thailand. Prepare your counter strategy for these requests before getting to the negotiating table.

TURKEY

Free-market policies adopted in the 1980s have moved Turkey from being a fringe economy to one of the most thriving of the emerging markets. Recently rebuffed by the European Union, the country still remains a major supplier to the global marketplace. Turkey is now a service economy (63 percent of GDP), but the economy continues to support substantial manufacturing and textile industries. There are still a substantial number of large state-owned companies (KITs) that put a drag on growth as they operate in subsidised, protected industries. Istanbul is a major transportation link for trade throughout Europe, the Middle East and the Asian subcontinent. Investors continue to be attracted to the educated workforce and strategic locale Turkey offers.

NEGOTIATION CULTURE

The Turks are known for their hospitality. Foreigners are often taken aback by the extent of their hosts' generosity. Experienced negotiators may see this outpouring of goodwill as a manipulative technique, and to a degree, it is. However, it can be taken at face value as part of a long-standing Turkish custom. Punctuality is a considered standard operating procedure, and its absence is considered a sign of disrespect. Make appointments well in advance of travel and confirm upon arrival in-country. This common courtesy can make or break early negotiation sessions.

Like much of commerce in this part of the world, haggling is commonplace, and nothing has a set price. If you're selling, start high, almost absurdly so and work back down. When buying, start at the bottom and begrudge every move upward. Always give an impression that you can walk away at any time. If the Turkish counterpart sees that your needs exceed your wants, this weakness will be immediately exploited. Much general conversation precedes business discussion, and there's a Turkish dislike of "cutting to the chase" just for the sake of expediency. Business moves at a slow pace; the Turks don't like to be rushed, especially by foreigners. Even when you've come to buy with cash in hand, expect a lot of chit-chats before a deal is cut. Due to its location, Turkey is chock full of polyglots. English, French and German are widely spoken, and translators for most European and Asian languages can be easily located.

Contracts will be signed in the local dialects, and all business materials should be translated for presentation. Contracts are usually stated in general terms with the specifics being hammered out over the extent of the relationship. Any insistence that quality control is made part of the contract will be met with considerable resistance. Counterparts will consider it insulting that a foreigner believes a Turk would deliver anything but the highest standard. Don't belabour the point. Western business dress is favoured among executives, and it's always best to remain conservative in appearance. Overly casual or exotic dressers will not be taken seriously regardless of wealth or title.

Temperatures run the gamut in Turkey; visitors should do meteorological research prior to negotiations. Home and family are sacred in Turkey and to be invited to a counterpart's home is a great honour. Never refuse this invitation unless it's impossible to comply; it's not offered lightly, and a casual refusal is a grave insult. While there's a good deal of socialising, don't expect wild after-work partying. Alcohol isn't always offered, as this is an Islamic, albeit secular, country.

UNITED KINGDOM

The United Kingdom is a dealing economy of the European Union. However, high growth and low unemployment (by European Standard at least) seem to vindicate the U.K.'s standoff approach to the European Union's single-currency policy. The United Kingdom has been a top player in international trade for centuries–it was certainly it first global practitioner ("The sun never set on the British Empire"). It also attracts an enormous amount of foreign investment to its own shore for high-tech, automotive and food processing. The country supports considerable financial services, insurance and transportation industries. Few international companies would consider their reach total without some form of U.K. presence.

NEGOTIATION CULTURE

The British are old hands at international business; their history of negotiation in that area goes back centuries. The depth of their knowledge is without comparison. Arrive in the U.K. thoroughly prepared and equipped with numerous options. Don't attempt to learn how things work at the table or your British counterpart will hand you your head. British business moves at a more deliberate pace than American business. Presentations should be detailed, with a minimum of hyperbole. The British have seen everything under the sun, and there's nothing new there, so get to your point.

The class system is still very much alive, and proper connections will make a difference for long-term projects. Government agencies and industry associations are good starting points for small-and medium-sized deals. Larger projects may require social introductions. Britain is an orderly society and punctuality is mandatory. Arrange appointments in advance and present in agenda as early in the process as possible. The British side will insist on having one, so it's to get your version in first.

Start your bargaining at a point only slightly distance from your projected goal. You can leave yourself some negotiating room but don't be excessive. Your British counterpart will have already researched the true value range of the deal. Finance is a major sector of the British economy; you should have no problem exploring options. British manufactured goods are generally high-quality, and there's little need to build provisions on that matter into the contract. Transportation and delivery requirements should be stated clearly.

The British tend towards detailed contracts littered with legal lingo. They may also insist upon having the contract bound under British law, although most international disputes on large matters will be settled in Brussels. If you're not from a society bound up with commercial contract law, be careful what you sign. Contracts in the U.K. are very binding, and penalties can be severe. Even in the post-Thatcherite U.K., unions are still very strong. Joint ventures should be heavily researched to uncover the possible effects of respective union involvement. Political intervention in this area isn't unusual.

The business lunch has been institutionalised in Britain; much negotiating will be done with knife and fork in hand. The British can be exceedingly charming and their manners put the world to shame. Enjoy the surrounding but resist the charm. It's all just so much posturing—usually perfect.

UNITED STATES

The United States is the largest economy in the world with its nearest competitors, Japan and China, being less than half its size. Low inflation, low-interest rates and low unemployment have stymied economists and thrilled foreign investors. The United States dollar is still the most widely used currency in the world and like it or not, every country measures itself against an "American" yardstick—another blow to the metric system. Its companies have some of the most commonly recognised international brand names, and its entertainment industry dominates world culture. The ability of the American economy to quickly recover from even the most substantial blows makes it a top choice for FDI even during the worst periods. The United States has hollowed out its manufacturing sector in recent decades and moved solidity into service and high-tech.

NEGOTIATION CULTURE

The Americans have only recently gained a grudging respect for other economies in the international marketplace. Formerly labelled naïve, Americans now use this stereotype to their favour as unsuspecting counterparts approach the negotiating table. Their business schools are the best in the world, and their society combines aspects of every other one on the planet. You may not like their style but don't underestimate their acumen. They aren't number one by accident.

True to history, Americans believe in winning wars by accepting a few lost battles. US negotiators are extremely nimble and can change strategy and tactics during a ten-minute break. You better be able to respond in kind. They're the practitioners of the original Cowboy mentality, and an American company will usually send its negotiators into the field with an unusual amount of authority. They often assume that counterparts have similar authority and are very disappointed with "errand boys" masquerading as an executive.

Individualistic by nature, they can also be good team players, but only when they get to select the team. Americans have a blind spot in that they believe everyone else in the world wants to be like them. They also believe that all markets should be as open as theirs purports to be. Even when you're on your own turf, the Americans want to play by their own rules. If you're on their turf, their lawyers will take great care in laying out the rules for you.

The American style is very direct, and they try to demand the same from counterparts. To not do so is to be labelled deceitful. Americans want you to look them straight in the eye and lay it on the line. They love confrontation and are not subtle in their intimidation techniques. The Americans play for big money and they play for keeps. As they're fond of saying, "If you can't take the heat, stay out of the kitchen". US negotiators generally start off from a strong position (at least in their minds) and are quite miserly with concessions. They've learned patience over the years, mostly from dealing with the Japanese and can wait until the end of negotiations to concede major points if necessary. Americans do, however, prefer speedy negotiations and chafe under too much extraneous socialising or postponement. They used to cut deals early (and to their disadvantage) just to save time. Nowadays they can afford to just leave.

VIETNAM

The Socialist Republic of Vietnam is one of the few remaining economies that are controlled by a communist leadership. Its *doi moi* policy, established in the mid-1980s, was designed to remedy the shortcoming of the centrally planned, profitless commercial sector. Foreign investment, both private and government subsidised, flooded in during the early 1990s but had slowed to a trickle as the less than stellar results have been tallied. The country has a formidable bureaucracy and corruption is rampant in both the government and private sectors. Vietnam's workforce is still predominantly agricultural, and it is one of the world's top rice exporters. The government has licensed numerous automotive companies to operate assembly plants, and some inroads have been made in developing light industry and technology. Tourism and agriculture remain the country's top sources of foreign currency. The local currency, the dong, is inconvertible on the open market.

NEGOTIATION CULTURE

The Vietnamese are accustomed to exercising enormous patience, something they perceive others especially Westerners, as being unable to do. "Time is money" isn't a Vietnamese concept; they're patient because the current pace of their culture makes patience feasible. Make sure that you're not locked into a very tight schedule. It will only work against you. The Vietnamese will often change agreed-upon terms overnight and seemingly arbitrarily, as a way of shifting the balance in their favour. Since there is little commercial law to enforce contracts, make sure that only the minimum amount of capital necessary for a project is turned over to the Vietnamese at any one time. It's not advisable to extend unsecured credit to a Vietnamese partner. Until recently, the Vietnamese were able to play one anxious suitor off against another. But an investment downswing has tipped the scales in favour of foreign capital. Let it be known that your company is willing to do business, but only in areas that show promise of investment returns within a reasonable period. This may be the most effective negotiation tool available to you. Many terms will be left unspecified by the Vietnamese, especially if they feel it's not to their advantage to clarify them. Your attempts to obtain a specific response will be met with vague nods or rapid changes of subject. Don't sign anything until the contract terms are transparent. It's hard enough to enforce a contract in Vietnam without adding vagueness to the procedure. Bureaucratic red tape is often used as an excuse for delays. The easiest way to deal with this is to play off the competitive spirit of the Vietnamese. Make it known early on in your discussions that your company will be pursuing many other projects within the country, that your appointments are numerous, and your agenda is organised. Any business that can't be completed in the time allotted will be postponed indefinitely. Most delays will evaporate when the spectre of competition enters the negotiation.

But make sure that you can back up your claims. The Vietnamese will generally hire their own interpreter. In many cases, this is a necessity, as the primary decision maker will not be fluent in your language. If the negotiations aren't proceeding in their favour, however, the Vietnamese will often claim that most of the problems are the result of linguistic or cultural misunderstanding. Bringing your own interpreter can counter this tactic.

Sl. No.	Country	Business Environment	Common Behaviour	Notes for Negotiators
1	Belgium	• Engineering and Shipping are the major Industries. • Hybrid of Dutch and French culture.	• Practical • Strict • Technical • Departmental	• First names are rarely used. • Factual information. • Can afford to say NO to just about any deal. • Be punctual and set-up meeting in advance. • Business visitors are expected to enjoy social aspects like Beer consumption.
2	France	• Business is rarely conducted with a sense of urgency • Personal relationships are important.	• Generous • Intimidating • Social • Hierarchical	• Well-read and cosmopolitan counterparts are appreciated • Proper use of language is a sensitive cultural issue • Rushing is considered as vulgar, "getting back" will not do • Avoid direct confrontation, love debate, not criticism.
3	Germany	• Expensive labour force, highly unionised and productive • Moving manufacturing to overseas to stay competitive.	• Impassive • Intimidating • Technical	• Codes and regulations dominate business • Contracts once signed strictly adhered by parties • Make straightforward presentation, time is important • Little regard for people who have to get back • Don't criticise your competitors, each one is judged on its own merit without comparison.
4	Ireland	• Service-based industries 67% of GDP • Low corporate tax.	• Compliant • Secretive • Social • Horizontal	• Known for their shrewdness, always ready to walk away. • Well read and will know about you before the meeting. • Present upside first but do share downside as well. • Highly technical presentation will be appreciated.

Sl. No.	Country	Business Environment	Common Behaviour	Notes for Negotiators
5	Israel	• A leader in food processing and diamond cutting • Government is protective for domestic enterprises • Foreign investors are rarely permitted for controlling stake unless guarantees of technology transfer.	• Aggressive • Exploitative • Intimidating • Horizontal	• Not generally bound be any rules • Often disregard the requirement of the counterpart • Punctuality and deadlines are demanded of foreigners but not of themselves • Little socialising attached to the business, expect no sugar coating • Proactive negotiator and like to anticipate their opponent moves, unveil your plans slowly • The brisk pace and quick YES or NO discussion.
6	Italy	• Frequent bouts with inflation, public debt and high government spending • Auto, clothing and jewellery is a major industry • Used to have trade surplus 10% to12%.	• Social • Compliant • Exploitative • Consensus • Divide and conquer	• Time is not an essence, but punctuality is appreciated • Connections are very important – personal references • Research and relationship builds most moves • Expressing a genuine appreciation of art, music, food and fashion will keep in good stead with locals • Social events and lengthy meals are part of business protocols • Can be argumentative about virtually everything, raising of voice and flailing hand motions are just passionate discourse not sign of anger.
7	Poland	• Strong Iron and Steel sector as well as machine manufacturing industry. • Import fuel and foodstuff.	• Stubborn • Social • Intimidating • Departmental	• Don't start negotiation unless you're serious • Negotiators will be pressed for technology transfer • The decision-making process is slow, communication is rigidly hierarchical • Be punctual and be prepared, socialising especially drinking is part of the business.
8	Russia	• Solid predictions of long-term, high single-digit growth are attracting investments • Has seen a steady rise of the mafiya in business.	• Aggressive • Deceptive • Intimidating • Hierarchical • Brinkmanship	• Send your best negotiator; graft is as common as taxation. • Remember that Russia needs practically everything. • Enjoy yourself but separate business from pleasure. • The most difficult discussion would be How buyer pays the seller. • Numerous trip will be required for med. to large ventures.

(Continued)

Sl. No.	Country	Business Environment	Common Behaviour	Notes for Negotiators
9	Sweden	• Major industries are steel, automotive, precision machinery and wood products • Dependent on foreign oil and foodstuff.	• Pragmatic • Social • Technical • Horizontal • Impassive	• Make an appointment well in advance and be punctual • Keep proposal realistic and be prepared for loopholes • Very formal and law-abiding society • Often delegate a good deal of decision-making to lower management • Swedish will most likely request to see the entire proposal before coming to negotiating table for efficient use of time.
10	Turkey	• Free-market policies have made them the global supplier • Service economy 63% of GDP • Large nos. of state-owned subsidised and protected Ind.	• Aggressive • Exploitative • Indulgent • Hierarchical • Social	• Known for their hospitality, which can be manipulated later • Punctuality is part of SOP and its absence is disrespect, this common courtesy can make or break negotiation sessions • Haggling is common nothing has set price • Always give an impression that you can walk away anytime • Business moves at slow pace, expect a lot of chit-chat • Home and family are sacred; never refuse such invitation unless it is absolutely impossible to comply.
11	UK	• Top player in International trade • Country supports considerable financial services, insurance and transportation industries.	• Financial • Stern • Arrogant • Legalistic • Hierarchical	• Be prepared and equip with numerous options. • Proper connections will make difference in long-term project. • Punctuality is mandatory, agenda based discussion. • Start bargaining when slightly distance from goal. • Contract in the UK are very binding and penalties can be severe. • Business lunch has been institutionalised, much negotiations happens with knife and fork in hand.

Sl. No.	Country	Business Environment	Common Behaviour	Notes for Negotiators
1	Egypt	• Textile and Food Processing and Tourism are major industries. • Bureaucracy and rubber stamping is part of the system.	• Aggressive • Secretive • Exploitative • Self-Righteous	• Sharp haggler is greatly respected • Lateness and postponements aren't unusual • Have solid agenda agreed upon before negotiations • The content of the contract can be re-negotiated many times throughout the length of the relationship.
2	India	• Most of the industries are open to foreign investment • The public sector is bureaucratic • Key industries are Software, Agriculture, Textile, etc. • Personal relationships are important for business.	• Social • Deceptive • Compliant • Departmental • Consensus	• English is widely spoken and is the business language. • Avoid use of the left hand for passing food items. • Not necessary good relations with one individual access to a source of power. • Direct saying NO is thought to be rude. • Don't assume strategies successful in one city will work in another.
3	Saudi Arabia	• Oil is a major industry • Depends on foreign firms for its infrastructure and construction.	• Impassive • Arrogant • Hierarchical • Stern	• Your punctuality will be taken as a sign of respect; counterpart lateness is a sign of power. • Business is conducted mostly on a person-to-person basis. • Almost take-it-or-leave it to approach towards foreigners. • Bargaining usually starts at an artificially high level when the Saudis are selling and below ground level when buying. • Generally bringing female executives to the negotiating table may spell disaster.
4	South Africa	• Rich with mineral resources with a bright future for shipping.	• Pragmatic • Stern • Social • Self-righteous	• Don't discuss racial topics. • All negotiations should clear benefits to both sides from completion of the deal. • Like straight talks and low-key sales approach. • Socialising is a big part of the business protocol.

Sl. No.	Country	Business Environment	Common Behaviour	Notes for Negotiators
1	China	• Commercial law favours domestic companies • Cheap labour rates • Shaky banking sector • A leader never delivers "bad news".	• Deceptive • Tolerant • Social • Exploitative • Hierarchical • Impassive	• Write explicit quality requirements • Friendship is a common ploy to secure concessions • Never use translator supplied by Chinese counterpart • Take notes and clarify inconsistencies across the table • Patience is greatly appreciated • Never reveal proprietary information until the deal is completed
2	Indonesia	• Government is a mandatory player in large projects • Key industries are petroleum, tourism, textile and mining.	• Deceptive • Compliant • Brinkmanship • Lenient	• Considerable haggling over even smallest point • Prone to Hierarchy, keep it simple • Contract is simply a guideline, don't assume it will be adhered strictly • Keep relationship healthy, get along or go home • Bureaucracy is a deep in government; "Consultation fees" is a form of getting things done.
3	Japan	• Moving to a competitive market economy • More mature market less insular • Cost of capital is very low so afford to take a long-term approach.	• Consensus • Impassive • Undecided • Divide and Conquer • Deceptive	• Group negotiations • No. of informal trade barriers and restrictions on FI • Formal and Hierarchy-based meetings • The person doing most talking won't be person-in-charge • Will discuss points as only part of pre-arranged agenda • Ambiguous responses–JIJO HENKO clause–permits complete re-negotiation if, circumstances changes.
4	Malaysia	• Strong agricultural, petroleum and rubber manufacturing. • Attractive economy for investors.	• Aggressive • Impassive • Ambivalent • Consensus • Hierarchical	• Detail-oriented people • Consensus decision-making and hierarchical communication • Most business discussion will involve some form of socialising • Reciprocations are expected, must keep the relationship strong and expect to succeed.

Sl. No.	Country	Business Environment	Common Behaviour	Notes for Negotiators
5	Philippines	• Build-Operate-Transfer-style infrastructure investment • Creating Export processing zones.	• Compliant • Secretive • Social • Indulgent • Deceptive	• Strong believe in forging and maintaining relationship • Acceptance of a favour will call for larger one in return • Gossip is both hobby and a business weapon • Law of all kind are very flexible, applied in direct proportion to the mount at stake • Leave your moral baggage at home and pack extra savvy.
6	Singapore	• Strict civil and Criminal laws • Hi-tech and financial service hub, shipping and transportation powerhouse.	• Aggressive • Impassive • Secretive • Exploitative	• Pay close attention to details, get everything in writing. • Don't give much emphasis on the value of personal relations. • Set a comfortable pace and stick to it, if pressed for time. • Quote, are normally higher than what they expect to get. • Negotiate with many options, expect hammering. • Misinterpretations are a standard way to cover mistake–tactic.
7	South Korea	• Striving to become a global player • Moving from industrial to technology base.	• Aggressive • Intimidating • Secretive • Deceptive • Divide and Conquer	• Will go toe-to-toe with anyone regardless of size • Diplomacy must be followed, don't be direct • May often make emotional pleas as negotiating style • Respect hardline strong opponents, Koreans always negotiate in teams, it will be exhausting.
8	Taiwan	• Industrial and technical power • High labour rates • Highly schooled in international business.	• Aggressive • Compliant • Secretive • Social • Hierarchical • Indulgent	• Learn everything about your counterpart but be discreet. • When they are selling–prepare for a healthy dose of indulgence; come prepared to the negotiation table. • Avoid fishing expedition, as they are serious in business. • Bigger concessions are saved for the end. • Great hospitality will be extended to put you in debt for easy negotiations and make you psychologically dependent, enjoy the ride but stay on your toes.

(Continued)

Sl. No.	Country	Business Environment	Common Behaviour	Notes for Negotiators
9	Thailand	• Textile, agricultural and aquiculture Ind. based. • Political chaos has affected FDI.	• Aggressive • Social • Exploitative • Indulgent • Consensus	• Personal relationship is an important part of doing business • Patience and punctuality is highly regarded • Avoid discussion/critiques in context to Monarchy • Request for "Tea money" or consulting fees are part of the business, prepare your counter strategy for such a request.
10	Vietnam	• Foreign investment both private and government are subsidised • Tourism, agriculture and automobiles are major industries • Local currency DONG is inconvertible in open market.	• Aggressive • Secretive • Deceptive • Hierarchical • Consensus	• "Time is money" isn't Vietnamese concept, so don't get locked into a tight schedule. • Will often change agreed-upon terms overnight, as a way of shifting the balance in their favour. • Little commercial law to enforce contracts. • Not advisable to extend unsecured credit to Vietnamese partner. • Don't sign anything until contract terms are transparent. • Bring your own interpreter can counter many misunderstandings which are used as tactic, as primary decision maker will not be fluent in foreign language.
11	Australia	• Well educated nation • Labour unions are strong • Mining and manufacturing are major industries.	• Aggressive • Realistic • Generous • Social	• Be Direct and avoid deception–Hands-on • Show up on time and come prepared • Avoid appearance of taking control • Contract will be written, detailed and enforceable • Encourage long-term relationships and work with people they count as friends.

Internal Negotiations–Building Stakeholder Alignment and Managing Expectations

Internal negotiations are often more difficult than negotiations with outsiders, it requires detailed planning to assess leadership as well as cross functions expectations in order to ensure their alignment. Building stakeholder alignment to improve the process and have consensus requires a lot of internal negotiations. There are two main reasons internal negotiations are more difficult:

a. Assumptions people make about internal negotiations.

b. Lack of preparation for internal negotiations, and lack of a common, internal process.

Assumptions are those things negotiators believe to be true without having tested their validity. Too often negotiators assume that because they work for the same organisation, "all are on the same team" and have the same overall goals, internal negotiations will be easy. However, within a single department, let alone across an entire organisation, there are likely to be conflicting goals, conflicting priorities and conflicting ways of being rewarded. When that gets coupled with a lack of direct authority over many of the people one must negotiate with, internal negotiations can be quite challenging. Even if a department or individual within your company shares your goals, they may not share your priorities. Or if they share your goals and priorities, competing commitments may mean that he/she can't honour the schedule you want for the negotiations. A negotiator may assume that negotiation goals and strategy have been communicated internally when often they have not. A negotiator must see his first task as gaining and building internal alignment about the goals, strategy and desired outcomes for any major negotiation by testing assumptions with key stakeholders.

When negotiating with someone outside the company, organisations and individuals often spend a great deal of time preparing for those negotiations. A salesperson might spend weeks or months preparing to negotiate with a major customer. A union president will spend months mapping strategy for upcoming contract negotiations. However, when an individual gets a call to meet with the boss to discuss upcoming negotiations, rarely will they see that meeting as a negotiation in itself. Thus, they often walk in unprepared or prepared to sell their answer or strategy without understanding the boss's interests, concerns and goals. We would argue that before developing a strategy, an effective negotiator must identify the key internal and external stakeholders and begin the negotiation process by sitting down with them one-on-one or in small groups to ask their interests, and brainstorm possible options for the upcoming negotiations. There is no common process, for how one goes about preparing for, conducting and reviewing negotiations. Too often, that is left to the initiative and skill of the individual who will be doing the negotiations; thus each person or department does its own thing. No organisational process or structure is in place to ensure thorough preparation for the negotiations or that a strategic review of the results occurs. No dissections of failed negotiations are done and the lessons from failures and successes are rarely recorded or shared with the entire team. There is no process to ensure that the organisation transfers the skills and tools of effective negotiation to its experienced or new employees. There is no common vocabulary or process for conducting negotiations. The two people on the same team may not only have very different expectations and level of skill in the negotiation process, but they also may not even speak the same negotiation language.

The key reasons for the failure of internal negotiations are as follows;

- Absence of background research or going too informally.

- Failure to assess stakeholders conflicting interest.

- Perceptions and rumours about one and another.

- Absence of belief that value creation is equally important in internal negotiations as well.

- Zero BATNA approach (Best Alternative to Negotiated Agreement).

- Failure to reinvent the relationship with colleagues and seniors.

Organisations must build a negotiation infrastructure that creates a common language and process for negotiations, and a common process and structure for conducting, reviewing and learning from their negotiations. Otherwise enormous amounts of time and money will be wasted, and valuable lessons and knowledge will go unshared. Thus, it is true that successful internal negotiations are key to successful external negotiations. Bear in mind Maureen Dowd's wise words, *"The minute you settle for less than you deserve, you get even less than you settle for."*

Sourcing Strategy–Tool for Business Transformation

Strategic sourcing can be defined as a process of selecting right supplier, i.e., supplier that offers the greatest overall benefit to the organisation, considering all other sourcing elements, i.e., right price, right time, right place, right quality, right quantity, etc. It is not necessarily the cheapest or highest quality supplier. A reduced cost of doing business includes not just a lower price, but also other efficiencies like inventory reduction, cost reduction, logistics cost reduction and ongoing productivity improvements that will positively impact the organisation bottom line/profitability. In short, strategic sourcing is driving for the highest value at the lowest cost.

Formation of sourcing strategy in any organisation is primarily responsible for catering organisation needs towards;

1. Cost Competitiveness

2. Customer Delight, i.e., quality, time, etc.

3. Performance and Innovation.

In most of the organisation, sourcing function is responsible for driving cost savings, i.e., either cost reduction through an innovative process or through creating competition in the marketplace as raw materials contribute almost 40% to 50% of the total cost of goods sold. Therefore, the organisation looks at their sourcing function to ensure that the strategies they follow give continuous cost improvement to their business. Customer focus approach is another criterion for ensuring that our sourcing strategies are leading towards the right direction. Quality, time and flexibility in our approach are the key tools to keep our strategies customer-focused. Supplier performance and supplier involvement are the other integral parts of our strategy formation exercise which can't be ignored.

Let's review first what to expect from purchasing function in an organisation, which needs to be addressed while forming sourcing strategies.

- Identify the right need.

- Avoid monopoly or sole-source situations.

- Identify potential competitive suppliers.

- Secure competitiveness.

- Consolidate volumes and leverage the same in sourcing decisions.

- Manage supplier's integration.

- Identify and cover risks.

- Show the economic impact of the technical decision.

- Frame contracts and agreements.

It is the right of the business to expect from their sourcing function that they get the best deal available from the market, steer the supplier selection process to ensure for a sustainable and mutually beneficial business relationship. Use cost analysis break down for negotiation, reduce switching cost to business and monitor and manage supplier performance and development.

The other major factor required to be studied for designing organisation sourcing strategy is supplier market analysis. It is important to define the market organisation in terms of numbers of the supplier is the marketplace, i.e., monopoly, oligopoly, open competition and geographic location of major suppliers. Supplier market analysis also evaluates attributes like production capacity (worldwide, per country, per region), who are top 10 suppliers, are companies over or under capacity, what is the capacity increase trend as well as demand trend of the product in other industries, what are the available and future technologies, in what phase of the life cycle curve are them, are all suppliers offering similar quality level, what are the major strategic issues in the market, which are the key success factors in this market, who are the other major purchasing players in the market, what are the main cost drivers in the market. One can get all this information by talking to few major suppliers on the issues discussed and can further cross-check that information with other. Nowadays, there are many specialised websites or publications and professional associations available to extend their support. One can also participate in external meetings and seminars to develop external industry network in order to use the same later whenever required.

Purchasing's Involvement Allows You to Focus On Your Core Competency.

You have a very important role in the organisation. Your expertise in your function makes you valuable. With Purchasing handling your procurement activities, you'll be able to spend more of your time on what you do best.

Purchasing's Involvement Helps You Avoid Last Minute Crises.

Your department is very busy with many competing priorities. In many departments that meet that same description, procurement activities are often put off until the last minute. This results in failure to find the best value in the market, paying expediting shipping charges, or worst of all, not obtaining goods and services on-time. Purchasing can help you avoid these headaches.

Purchasing's Involvement Gets the Most Out of Your Budget.

Unless your department invests in negotiation training for its staff and gives them the daily opportunity to negotiate with suppliers, suppliers may have an advantage in bargaining. Because the purchasing staff regularly receives negotiation training, negotiates daily and keeps up to date with the latest cost-saving techniques. Purchasing can help save your department money and alleviate some of your budget constraints.

Purchasing's Involvement Can Uncover Unforeseen Obstacles.

Whether it be seeing the warning signs of a supplier in financial trouble, identifying a material in short supply or just knowing the typical timelines associated in getting the goods or services you need, Purchasing reduces risks to your department's operations.

Purchasing Scenario and Potential Action

Let's discuss here below some possible Market Condition vs. Supply Situation vs. Possible Purchaser action in that context;

CASE # 1

Market Condition

There are multiple sources in the market and real competition happening.

Supply Situation

- The supplier does not believe there is competition, or you will move your business.

- The current supplier does not want to compete.

- The current supplier wants to keep the business.

- Other suppliers want to gain the business.

Purchaser Action

- Move all or part of the business to the alternate supplier and educate your supplier.

- Agree cost improvement principles/move the business.

- Threaten loss of business for improvement.

- Build the belief that new business is there to be won for a better deal.

CASE # 2

Market Condition

There are multiple sources in the market, but suppliers are not competing.

Supply Situation

- Supplier recognises the danger of competition.

- Buying pressure is weak.

- No threat from new materials or new technology or new source.

- Government or International regulatory restrictions.

Purchaser Action

- Create uncertainty for loss of business/focus on "Weak Link/Seek extra value".

- Commit to change/find non-competing allies.

- Be a champion of change/search for options/develop a new source.

- Build your understanding/search for loopholes/look for alternatives.

CASE # 3

Market Condition

There is only a single source and no real competition.

Supply Situation

- Supplier applies planned pressure.

- Supplier believes that there is a competition.

- Supplier believes that there is no real competition.

Purchaser Action

- Build your knowledge, bombard with logic.

- Threaten to use it. but recognise the danger.

- Develop if, possible/logical and emotional PPCA/liaise with R&D to investigate uses of different process and materials.

New Manifesto for Sourcing Strategy

Every company wants their top line to grow without compromising on the bottom line. A strategy that ignores global competitive sourcing opportunities may cost a business both its customer and vast sums of money. An organisation needs competent sourcing function to facilitate fair competition and to protect the company and its customers. If it is serious about getting their "Chemical Sourcing Act" together, it must look at more than just the price of the chemicals it buys and should go deeper while choosing supply-chain partners for a smooth and trouble-free relationship. Such a relation is only possible with the help of detailed techno-commercial and behavioural assessment.

Most advanced companies are switching from low-cost country sourcing to best cost country sourcing recognising that lead times are parts of the total cost of ownership. Poor sourcing and supplier competency are the main factors limiting productivity, profitability and growth of the organisation, particularly in companies where sourcing is country-centric, and they are not able to keep pace with fast-growing global competition. Sourcing chemicals are not like buying hardware or machines. It needs to establish uniformity among the chemicals we purchase, use and store. When we buy the same chemical from five different companies, we end up buying five different levels of quality. This all adds up to confusion for quality/process control functions for setting up different process standards in manufacturing and safety standards. Benchmarking requires prime supplier's attributes helps in long-term hassle-free supply-chain management.

Increasing customer demands, innovations in technologies, higher competition in the global environment, decreased governmental regulation and increased environmental consciousness have forced organisations to nurture their supply-chain and put more efforts into the supplier selection process. The fit between how organisations select suppliers and contingency variables changes and hence organisations need to adapt to the new level of the contingencies to avoid loss of performance.

The key and perhaps the most important process of the purchasing function is the efficient selection of suppliers because it brings significant productivity and cost improvement to the organisation. The objective of the supplier selection process is to reduce risk and maximise the total value for the buyer, which involves considering a series of strategic variables. Companies can realise significant material cost improvement when they successfully access and develop low-cost sources in emerging regions.

For any company, outsourcing means replacing the in-house production of an intermediate or final product with that of purchased or sourced from outside the company/country and is an integral part of value-added-chain. While the word global sourcing is more looked as cost improvement exercise by many, the other aspects which need attention are the substantial amount of risk and complexities, global sourcing adds to the value chain.

Following is the list of identified risks and complexities based on the survey reported in "Low-Cost Country Sourcing report 2006" presented in Chicago seminar*:

Risks and Complexities	% age
Poorly Developed Infrastructure	48
Complexity of Transport and Logistics Operations	45
Immature Suppliers	38
Trade Regulations, Customs and Tariffs	37
Cultural Differences	27
Government Regulations	23
Others	12
Unreliability of Quality	51
Unreliability of Delivery	48
Supply-Chain Safety and Security	39
Compromised Efficiency	27
Political Issues	26

As per data presented in a conference on Sourcing in Low-Cost Countries in Chicago on Sept.'06.

Companies can overcome these risks by forming deep, lasting relationships with vendors which help in ensuring intellectual property because business licence and value-added tax fraud can be common, companies should perform comprehensive documentation checks to ensure that licenses are proper, and taxes have been paid. Technical issues like poor quality, low-tech, infrastructure weakness can be handled by providing suppliers with the latest high-tech equipment and develop the ability to solve problems on-site. Efforts in training and developing staff, hiring good local managers, giving appropriate performance-related compensation and incentives, ensuring mature process help in building a capable local organisation to handle risks associated with poor-quality staff, high employee turnover, lack of experience, etc.

Establishing a new manufacturing source brings change that affects the supply-chain, especially early in the process with risks like difficulty in communication with staff and vendors, availability of baseline data, understanding time requirement to complete transition to the low-cost country source.

To reduce or eliminate these risks, companies can establish a rigorous product development process with early involvement of emerging region suppliers into the product development process.

Significant supply-chain disruption can reduce the company's revenue, cut into market share, inflate costs, threaten production and operation, run over-budget. Such disruption also damages credibility with investors and other stakeholders. With operations scattered around the globe, companies face a host of new perils: political and currency risks, cyber-attacks, failed communications with suppliers, just-in-time delivery strategies and dramatic, unpredictable risks associated with terrorism. Companies also still face traditional property related risks to their supply-chains such as fire, natural disasters, power grid blackouts and equipment breakdowns.

In a study of more than 800 companies that announced supply-chain disruption between 1989 and 2000, Singhal and Hendricks found that ill effects of a supply-chain disruption don't disappear quickly. Changes in operating income, sales, total costs and inventories all remained negative for the problem companies in the two years after their problems were disclosed*. Like a heart attack that cuts off the flow of blood, a supply-chain glitch cuts off the flow of information or supplies and similar to heart attack, it has lasting effects on a company's health.

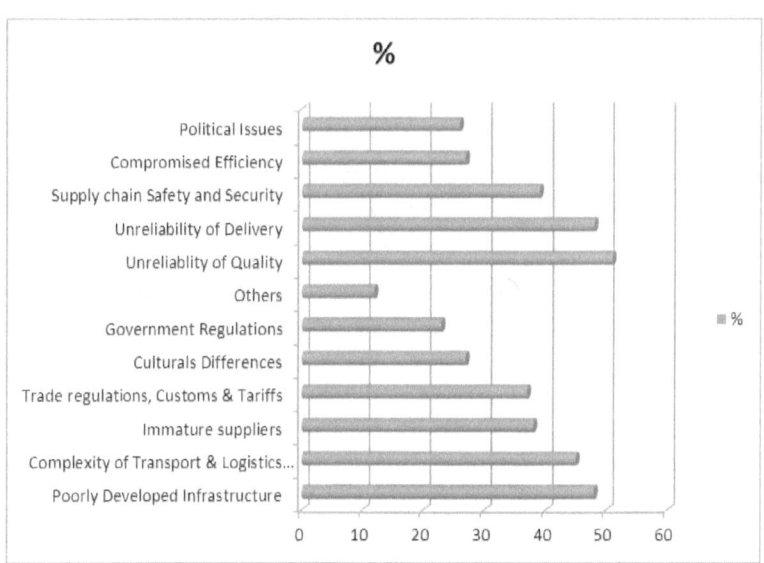

*Hendricks, K. B., V. R. Singhal. 2003b. "An Empirical Analysis of the Effect of Supply-Chain Disruptions on Operating Performance." Working Paper, DuPree College of Management, Georgia Institute of Technology, Atlanta, GA.

Global supply-chain forms the backbone of the global economy, fuelling trade, consumption and economic growth, trends such as globalisation, lean processes and the geographical concentration of production have made supply-chain network more efficient, but have also changed their risk profile. Most organisations have risk management protocols that can address localised disruptions. However, recent high-profile events have highlighted how risks outside the control of individual organisations can have cascading and unintended consequences that can't be mitigated by one organisation alone.

The government also been increasingly challenged to understand and manage risk across the global supply-chain. The political, economic and security implications of regulating in a complex environment have necessitated new approaches for public-private collaboration. However, prediction of specific disruptions is felt to be less important than having the resiliency in place for effective response, no matter what the cause. Disruptions are categorised under four broad groups: Environmental, Geographical, Economic and Technological.

Environmental risks refer to events such as volcano eruption, earthquake, tsunami, floods like natural disasters are most likely to cause systematic supply-chain disruption. Geographical disruption encompasses a range of potential disruptions including conflict and unrest, terrorism, organised crime and corruption. Economic disruption covers a range of issues including currency fluctuations, commodity price volatility, sudden demand shocks, border delays and ownership/investment restrictions whereas, technological risks are associated with information/communication disruptions and infrastructure failure. Few of such risks and its definitions are as follows;

- **Border Delays** – Delays and restrictions on the free movement of goods and people across international borders due to screening programmes, customs clearance or immigration controls.

- **Commodity Price Volatility** – Severe price fluctuations that make critical commodities unaffordable, slow growth and increase global tensions.

- **Conflict and Political Unrest** – Military action or aggressive foreign or trade policies on the part of global or regional powers that disrupt political or social instability, negatively impacting investment and financial markets.

- **Corruption** – The abuse of power for personal gain by businesses and public officials that undermines the rule of laws, governance, investment flows and economic development.

- **Currency Fluctuations** – Global savings and investment imbalances that foster unsustainable current account imbalance, unsustainable levels of external debt and an ultimately wide swing in foreign exchange rates.

- **Energy Shortage** – The impact of the sudden change to the availability of energy, e.g., shortages in electricity supply.

- **Export/Import Restrictions** – Restrictions on the type and quantity of goods exported or imported within a specific country or countries by a government.

- **Extreme Weather** – Storms, cyclones and other acute weather events that cause harm to lives, human health, infrastructure, property, economic activity and the environment.

- **Illicit Trade and Organised Crime** – Unchecked spread of illegal trafficking of goods and people through the global economy, highly organised disciplined and deep-rooted global networks, committing criminal offences.

- **Information and Communication Disruptions** – Single point system vulnerabilities that trigger cascading failure of critical information infrastructure and networks.

- **Natural Disasters** – Earthquake, volcanic action and other geographical catastrophes that cause harm to lives, human health, infrastructure, property, economic activity and the environment.

- **Nuclear/Biological/Chemical Weapons** – The availability of nuclear, chemical, biological and radiological technologies and materials intended to cause harm.

- **Ownership/Investment Restrictions** – Barriers to market entry, e.g., restrictions on airline ownership and control, sabotage rights.

- **Pandemic** – The incidence and patterns of both known and emerging infectious diseases that shift to new regions and population segments through a series of pandemics or sub-pandemic outbreaks, threatening global health and economic activity.

- **Maritime Piracy** – The spread of violence or depredation on the high seas, directly impacting the global passage of goods and people.

- **Shortage of Labour** – A shortage of skilled and unskilled labour directly impacting the effectiveness of supply-chain and transport network.

- **Sudden Demand Shocks** – Sudden changes in demand across an industry or sector due to external events, e.g., immediate decline in airline passengers or decline in sales, etc.

- **Terrorism** – Individuals or non-state group that successfully inflicts large-scale human or material damage.

- **Transport Infrastructure Failure** – Cascading failure of critical transport infrastructure and networks due to frequent strikes or any other reasons.

- **Water Security** – Decline in the quality and quantity of fresh water combined with increased competition in the industry like food and energy production which needs a high amount of water.

As the above risks can only be sorted out by systematic approach and coordination between government and business community, it is important to consider these perspectives while selecting the location of a new supplier to support global operations. Here government should be known for improving international and inter-agency compatibility of resilience standard and programmes, whereas the business community should more explicitly assess supply-chain and transport risks as part of procurement management and governance processes.

Further, government and business world should jointly develop trusted network of suppliers, customers, competitors, improve network risk visibility as well as pre-and-post-event communication on systemic disruption and balance security and facilitation to bring a more balanced public and private sector discussion. These are macro-level assessment; the organisation should take care while deciding on new supplier development. This study is not covering such macro-level environmental/political and economic issues and focusing more on supplier specific aspects with respect to company goodwill, man, machine, cost and service level aspects.

Increasing focus on low-cost country sourcing has enhanced the supplier selection importance by many folds. On the other hand, growing consumer demands for sustainable products, despite worries to the contrary; but often with lower margins on such products and services ~ from the supply-chain perspective, this underlines the difficulties that exist in trying to adapt historical supply-chain approaches to suit new demands instead of wholly thinking the new approach. The primary reasons behind companies going for low-cost country sourcing are as follows:

Description	%
Lower material Cost	48
Lower Labour Cost	47
Pressure from customer to reduce prices	42
Close to Customer	11
Other	18

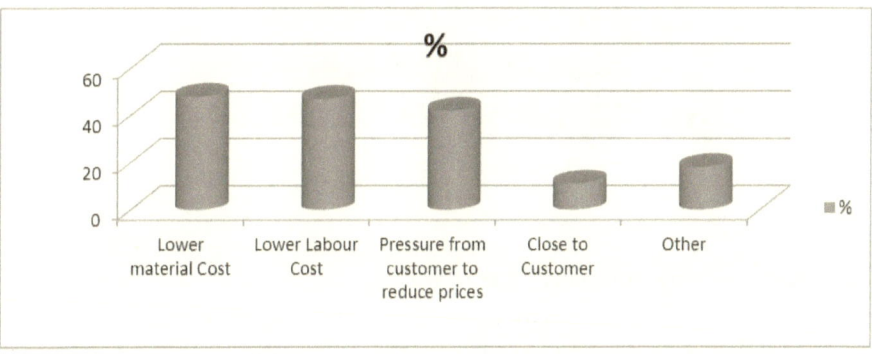

(As per data presented in a conference on Sourcing in Low-Cost Countries in Chicago on Sept.'06)

Who Sources What From Where

Low-Cost Source Countries	%
China	70
India	48
Other Asia	47
Eastern Europe	36
Other Latin America	22
Mexico	20
Other	13

(As per data presented in a conference on Sourcing in Low-Cost Countries in Chicago on Sept.'06)

While sourcing in low-cost countries may offer huge benefits, it also has many pitfalls and difficulties. The obvious difficulties are cultural and political differences. Companies seeking low-cost country sources for the first time may find that establishing and qualifying those sources and making the transition to them may take longer than expected.

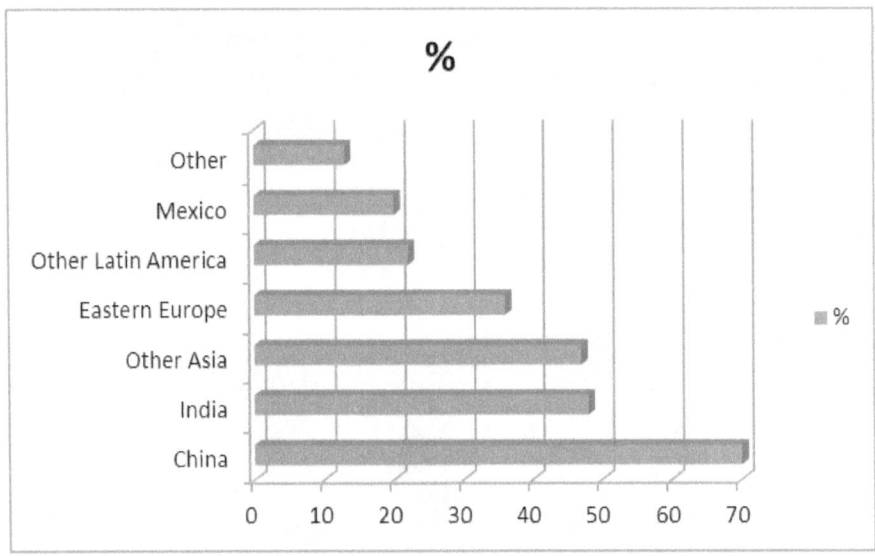

Therefore, for any company that is willing to outsource part of its value-added chain, it's important to ensure that their supplier selection process in the global sourcing function addresses all these risks and adequate provisions or processes are in place to safeguard the buyer's long-term business interests. For instance, detail working on overall storage and logistics-related cost involved with the supply-chain cost while comparing the sourcing cost from local source vs. overseas supplier, product price benefit vs. increase in inventory cost and increase in rejection risk, increase in lead time involved in the procurement process as well as substantial increase in overhead cost associated with additional infrastructure, people and other indirect cost involved with running these global sourcing offices outside the country.

Therefore, it becomes equally important for a foreign buyer to evolve a process of source selection in context to global sourcing which has provision to deal with intense global competition, sustainability of the relationship, thrust for continuous cost improvement, total quality excellence in order to make this new association win-win equation for seller as well as for buyer. Given the importance of the supplier, it would be reasonable to expect the appropriate resources and methodologies are employed in selecting the best suppliers.

It can also be described as finding out what an organisation wants to achieve by selecting a supplier so that they can refrain from dropping a supplier for the wrong reasons. In case the main motive of replacing an existing supplier is unsatisfactory supply performance, then there is no need to modify the specifications in the supplier selection process but to find out the new ways to ascertain performance-related aspects without getting into real business/transaction relationship. Which means customer should investigate supplier assertions on economies of scale, superior infrastructure and supplier procurement practice. It is also vital to verify the availability of specialised professional skills and the dynamic areas of quality and costs of staff in offshore locations.

SOURCING STRATEGY PROCESS

GLOBAL STRATEGIC SOURCING–PATH FOR BUSINESS TRANSFORMATION

While we explore global strategic sourcing, it is important to understand what the main characteristics of being strategic are with respect to sourcing function.

Spend Analysis: Strategic purchasing teams examine the amount of money they spend in each category of goods and services and use this analysis to identify opportunities for improvement.

Supplier Relationship Management: Strategic purchasing teams measure supplier performance and regularly spend time meeting with their most important suppliers to implement improvements.

Technology Implementation: Strategic purchasing teams frequently update and add technologies that measurably reduce costs, decrease cycle time and make the purchasing process more efficient.

Developing Project Plans: Strategic purchasing teams use project management practices to map out both recurring activities and one-time projects.

Enterprise-wide Contracts: Strategic purchasing teams consolidate spend across all parts of their organisations and enter into contracts with a limited supply base to serve the needs of the entire organisation.

Forecasting: Strategic purchasing teams regularly document changes that they foresee in price levels, availability and markets to ensure a competitive advantage for their organisations.

Involvement in Spec Development: Strategic purchasing teams are involved at the early stages of specification development, lending specialised knowledge in material availability, cost drivers, standard parts and reliability of supply.

Development of Productivity Tools: Strategic purchasing teams develop tools (e.g., RFP templates) so repetitive tasks can be done more quickly and error-free.

Supplier Development: Strategic purchasing teams don't blindly accept the suppliers and products that are currently available. They work with suppliers to develop new capabilities or products that will improve cost or quality.

Work Responsibility Refinement: Strategic purchasing teams constantly identify ways to automate, delegate or eliminate tactical, non-value-added work.

Further Our Global Sourcing Strategy Needs to Study the Following

Costs – A global sourcing strategy is often used to benefit from lower labour costs abroad. But there are also other additional costs for a buying organisation to bear that aren't part of domestic transactions. They include multi-model freight charges, broker fees, bank fees, taxes called duties and insurance to name a few.

Laws – Global sourcing forces buyers and suppliers to choose one of three bodies of law to apply to their contract: the law of the buyer's country, the law of the supplier's country, or one applicable under a treaty accepted by both countries.

Currency – The buyer and the seller must agree on a currency to use. While some buyers insist on their own currency for simplicity's sake, prudent decisions consider the use of the supplier's currency when the buyer's currency might strengthen relative to the supplier's currency between the agreement and payment dates.

Lead Time – Lead time for global purchases is usually significantly longer than for domestic ones. This is due to ocean travel being slower than air travel and customs clearance adding time not involved in domestic sourcing.

Language and Culture – If you're unfamiliar with the supplier's language and culture, you increase the risk of communication challenges, misunderstandings and offensive or uncomfortable encounters.

Transportation – While domestic sourcing usually involves one shipping mode, global sourcing involves multi-modal transportation—a strategy for combining air, water and ground transportation to get goods from the supplier to the port of the supplier's country to your country's port to your dock.

Payment Methods – Global sourcing often involves payment using a letter of credit which requires the involvement of both the buyer's and supplier's banks.

Because of the differences between domestic and global strategic sourcing, you need the right skills, knowledge and tools to successfully source globally. Here are five of the most important global strategic sourcing tools.

1. **A Business Case.** Many company executives need to be convinced that global strategic sourcing is a smart move. A business case documenting the researched savings potential is a tool for convincing them. Let's say that you get an answer of $30,000 savings, says Dick Locke, author of the book *Global Supply Management* "Is that something your company can afford to walk away from?" Only after management reviews a well-written business case can that type of question be answered.

2. **Cultural Research.** Learning about your suppliers' culture is critical for global strategic sourcing success for two reasons in Locke's opinion. One is that, if you have misunderstandings, they can get in the way of closing deals, he says. The second reason is that if you understand cultural differences, you might be able to be more demanding of a supplier in another country than you would be in your own country.

3. **A Landed Cost Model.** In global strategic sourcing, there are more cost components compared to domestic sourcing. A landed cost model helps you include global sourcing cost components such as multi-modal freight, duties, customs fees and more into your analysis.

4. **An INCOTERM'S Chart.** INCOTERMS define the responsibility for the buyer and seller in terms of handling and paying for customs and shipping, according to Locke, it also defines where the risk of loss transfers between the seller and the buyer. There are 13 INCOTERMS which differ from the F.O.B. terms used domestically to DAP and others used in international trade.

5. **A Transportation Time Chart.** The proximity of suppliers influences transportation times, which influences your inventory strategy. Always know your transportation times.

Literature review has shown the strategic changes in manufacturing firms and the greater roles played by suppliers in these firm's success. While Porter (1979) recognises the supplier as a competitive force that must be recognised, Prahalad & Hamel (1994) have identified selecting the right supplier as an opportunity to build on the firm's core competencies. It has been seen how successful firms such as Honda and Walmart have recognised the supplier as playing a key role in their success.

Jagdev & Browne (1998) defined the extended enterprise as cooperation on design, development and manufacturing across several independent manufacturing enterprises and suppliers. They have shown how the supplier's relationship has moved to having joint ownership for the satisfaction of their joint end customers; they maximise on their respective core competencies and maintain a fundamental level of trust. Childe (1998) brings this concept further when he describes the supplier as becoming part of the principal company and after some time become interdependent. An adversarial approach to dealing with suppliers offering short contracts for the lowest price has moved to long-term relationships where mutual success is important, and each participant capitalises on their respective core competencies.

As the role of the supplier moved from that of an external entity to being a strategic partner in the extended enterprises, the importance of selecting the right supplier has become much greater. When selecting new suppliers, it is important that firms ensure that the new supplier brings the appropriate competencies and is a good strategic and cultural fit before embarking on a long development or manufacturing project.

Much of the published literature on supplier selection has focused on mathematical models that may be used for decision-making rather than identifying the best criteria for supplier selection.

Weber (1991) has listed Dickson's 23 vendor selection criteria, first published in 1966 and examined their relevance in 1991. Through a literature review, he has rearranged them in new priorities based on the number of articles citing these criteria. It is worth noting that the relevance of price has increased significantly. The advent of JIT manufacturing is also very evident as Dickson's Geographical Location which we now call lead time has become more important due to attempts to reduce inventory. While other authors have published work on vendor selection criteria, nine are significantly different to those published by Dickson in 1966 and principally revolve around Price, Quality, Capability, Reliability of delivery. Through the review of material published on vendor selection criteria, little has been done to review business significance from a customer perspective and to offer weighted scorecard.

A review of ISO guidelines has done little to identify the critical criteria for vendor selection. These regulatory bodies have been clear in placing the onus of the firm that it is their responsibility to ensure supplier's ability to supply goods or services, to meet the appropriate requirements but they do not suggest how to. In vendor selection, these organisations have been most useful in their accreditation role. Many organisations identify these ISO and FDA/cGMP accreditations as a minimum quality requirement in their vendor selection criteria often using this as an excuse for not carrying out an audit at potential new vendors.

The past few years have seen a significant transition in the corporate view of supply-chain sustainability. In 2008, the Economist Intelligence Unit identified the supply-chain as the weakest link from a sustainability perspective, with only a few companies giving it much attention in comparison to other parts of the business. Historically, supply-chain sustainability was feared as a new cost and management burden, driven by vague social responsibility aims. Now there is a growing understanding that sustainability can bring wide-ranging benefits such as innovation, efficiencies, better risk management, improved brand value and even new revenue generation initiatives, this reflects the fact that supply-chain are often large, complex and difficult to change that's why it makes supplier selection process more critical and strategic in nature to meet short and long-term objectives of the business and delivering on this will be huge challenge for leading global companies planning to spread their supply-chain footprints in emerging regions.

Based on PwC survey, it's interesting to note that on average only 36% of manufacturing is being outsourced, indicating that function like manufacturing and sourcing remain cores of the supply-chain and keys to achieving tighter supply-chain integration. Due to global disasters in the past few years, some companies have actually brought some supply-chain activities back, close to home to reduce risks. Majority of the companies outsource half of their transportation and warehousing activities, regarding them as commodities that can be handled by a partner. But keep the customer order desk-in-house to maintain control over interaction with customers. Companies are still focusing on the basics of running a supply-chain in a cost-effective manner and delivering goods sufficiently well to satisfy their customers by concentrating on achieving continuous improvement in cost, lead time and waste reduction. These basis capabilities are simply pre-conditions for doing business and don't enable a company to outperform the market.

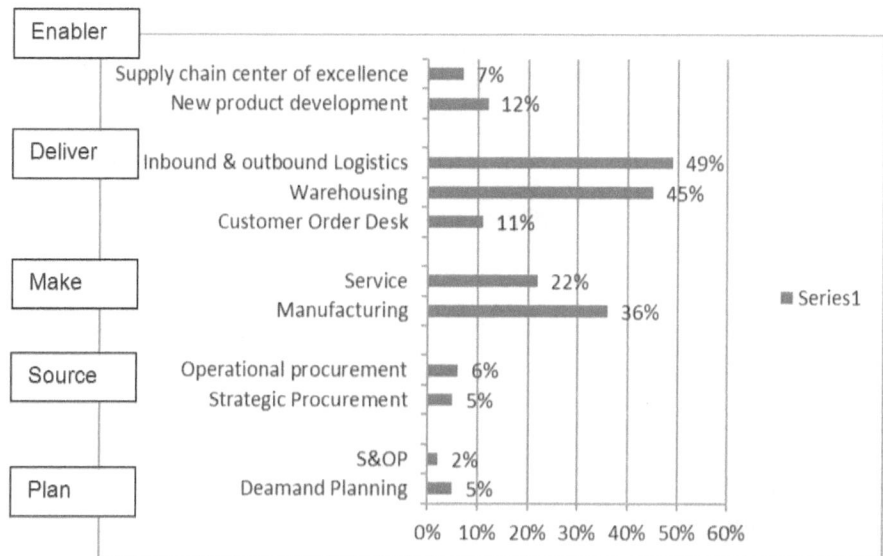

*PwC (2013) ~ Next generation supply-chains: Efficient, fast and tailored–Global Supply-Chain Survey

From chemical and process industry perspective, companies typically manage their planning, manufacturing, operational procurement and delivery function regionally and strategic procurement functions globally. Current trends say companies outsource about 5% of planning function, sourcing and enabling activities and 13% manufacturing activities and 7% to 45% of delivery activities. The most important value driver for the chemical industry is cost minimisation (87%), maximum delivery performance (87%) maximum volume flexibility and responsiveness (77%) and complexity management (72%). Companies focus on continuous improvements in production efficiencies and inventory management, coupled with process simplification to drive down costs and on end-to-end supply-chain planning and visibility.

Top differentiating practices in chemical and process industry with procurement and supply-chain functions are as follows:

- Inventory reduction.

- Decreased manufacturing cost through reduction of waste.

- Reduction in-process flow complexity.

- End-to-end supply-chain planning and visibility.

- Collaboration with key customers on planning, i.e., effective forecasting.

- Order fulfilment cycle time reduction to improve information flow.

- Internal capacity flexibility 80% to 120%.

- Involvement of partners for capacity reservation.

- Development of multi-skilled employees in order to cope with complexity.

- Making to order.

- Inventory policies distinguished by product family and storing location.

- Multiplication of source and sole-source avoidance.

- Visibility over short-term supply through order traceability, Vendor Managed Inventory.

- Visibility and regular monitoring of the main supplier's operational indicators.

- Responsible supply-chain partner footprint and procurement framework.

- Integrated risk management.

- Agreement of supply partner to adhere to highest ethical standards.

- Transfer pricing.

- Localisation of procurement organisation in tax efficient countries.

- Manufacturing optimisation through toll and contract manufacturing.

- Best cost country sourcing.

- Automation of processes in order to cope with complexity.

- Internal carbon footprint optimisation and improvement.

- Import/Export optimisation.

- Intellectual property and patent royalty optimisation.

Sourcing organisations should investigate not only the potential supplier's capabilities but also their commitments for providing satisfactory delivery performance. On time delivery, low commodity damage rates and delivery adjustment flexibility are the most important factors due to the long lead time in case of offshoring and hence on-time delivery is considered an important criteria.

Literature proposes that sourcing organisations could evaluate the fitness of supplier's customer support equation by examining following factors like accessibility, timeliness, responsiveness, dependability. The empirical findings indicate that service levels demanded by case organisations are largely depending on what type of products they normally outsource. Reliability, flexibility, consistency and long-term relationship are four significant new entrants into the list of critical success factors for supplier selection.

Based on the review of literature, it seems appropriate to conclude that supplier selection criteria will continue to change based on expanded definition of excellence to include traditional aspects of performance (quality, delivery, price, service) in addition to non-traditional, evolving ones (JIT, communication, process improvement, supply-chain management) to drive competitiveness throughout the global supply-chain allowing factors other than price to assume greater importance in the supplier selection problem.

Literature review also signifies how important it is to have a vendor assessment process in place, while this system will focus on risk and gaps with the particular vendor but also provide handholding to fill those gaps and mitigate the related risks. In general, it specifies what are the major benefits of assessing vendor risks and satisfaction level it leads to on sustainable basis, which can be well integrated into supplier selection process

Should You Dual Source or Single Source?

Dual source means to use two preferred suppliers to provide the same product or service. Single source means to use just one preferred supplier, despite there being multiple capable suppliers available.

Many purchasers decide to single or dual source prior to issuing an RFP or tender based on certain assumptions. Common assumptions are that there is a lower cost with a single source due to leveraging your volume but less risk with a dual source due to having a qualified supplier up and running if the other fails to perform.

Those assumptions may be right sometimes, but as a professional, you should make decisions on facts, not assumptions. To acquire facts, request three prices from your suppliers:

a.   For 100% of your business,

b.   For 70% of your business,

c.   For 30% of your business.

Upon bid receipt, compute the cost of doing business with the two qualified suppliers who bid the lowest for the 70% and 30% chunks of your business. Compare that cost with the lowest qualified single-source bid.

Is there a cost difference between the single and dual source options? If so, does the lower risk justify the premium?

DATA ANALYSIS ESTABLISHES RELATIONSHIPS AND TRENDS

Based on the survey with participants from Chemical industry in India and follow-up discussion, the following points are evident;

1.   For many organisations, these processes have been created only to meet commitments in their quality manual or quality requirements of independent accreditation agencies. As a result of this, the procedure has a strong quality focus with many criteria taken directly from ISO 9001:2000. This leads to a high dependency on independent industry standards rather than looking at the specific needs of the organisations.

2.   Conversations with many of the small and medium scale companies have indicated that the relationship with their suppliers is highly valued, and first impressions from face-to-face meetings generally supersede any formal analysis.

3.   Many of the companies have relied on the potential vendor to complete a questionnaire which may or may not be followed up with an audit or action plan.

4.   Most of the companies have some form of vendor management system or supplier performance card which uses measures such as % on OTTR (On time to receive) and % quality defects.

5. In almost all cases, criteria are attributes rather than variables that can be measured and compared. For example, the presence of an ISO system does not necessarily mean it has been effectively implemented and as a YES/NO result does not allow for comparison between two companies who both have ISO's system.

6. In highly regulated industries such as Pharma API's as well as Electronic chemicals production, the selection of vendors is constrained by independent accreditations and costly validation requirements. This reduces the scope to change supplier. Potential competitive advantages of cost, quality or lead time may be offset by the time and expense required to validate a new vendor.

In summary from the limited number of examples above, many organisations use their vendor selection processes as a stage gate to rubber-stamp the decision to add a new vendor to their Approved Vendor List. The processes are highly reliant on the individual needs of the organisation and do not follow any standard operating procedure for supplier selection.

All companies agree with the importance of having good suppliers and view their suppliers as business partners. Many have insisted that a face-to-face meeting is of greater value than the paper exercise of a vendor assessment form. It is evident from all of the examples above that the selection criterion which is in use lacks clear matrix and is generally flexible to meet the company's individual needs.

Strong supplier selection criteria and developing supplier involvement programme enhances communication and creates an environment of trust that builds a fertile relationship. These relationships improve performance by eliminating raw material stock-outs, increasing on-time delivery, reducing in-transit damage and improving incoming product quality and the firm will be able to meet its manufacturing objectives regarding production costs, WIP, inventory levels, product quality and on-time delivery to the final customer.

The multiple case studies indicate that for those SMEs who do not have any experience of outsourcing and also do not have much resources and time to dedicate in vendor selection, the cooperation with an intermediary is recommended. When companies have a well-thought plan with accumulated experience in outsourcing, they can shift from indirect sourcing to direct sourcing. SMEs should try to streamline the vendor selection processes and explore unbiased vendor selection methods.

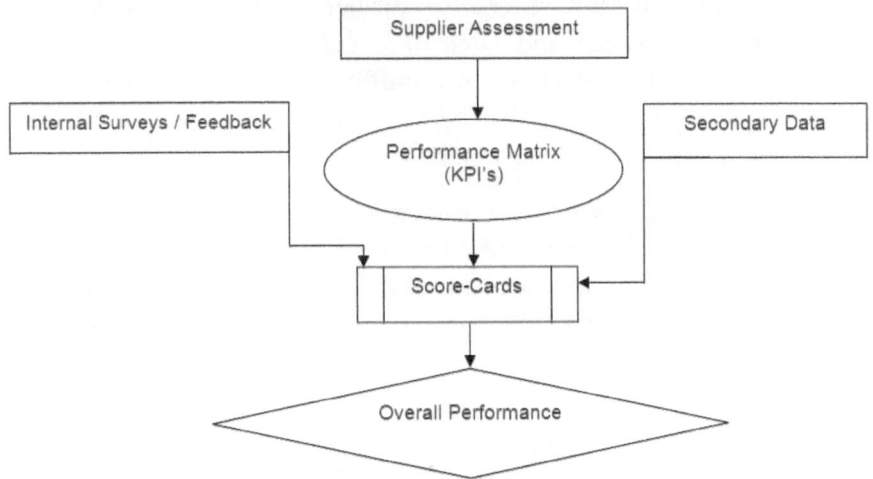

MONITOR AND ESTABLISH NEW BENCHMARK FOR CONTINUOUS IMPROVEMENT

As the acid test of the conducted survey outcome in the supplier selection process, a good approach would be to evaluate some new suppliers according to a proposed criterion and monitor their progress over a period against the earlier benchmark. As purchasing commands, a significant position in most organisations since purchased supplies, products and services typically represent 40% to 60% of the sales cost. This study would identify the best potential suppliers. This work would be very difficult to do retrospectively as the evaluation must take place at the beginning of the relationship. Subsequently, effective supplier's continuous development programmes can be leveraged, wherein it is reasonable to expect the vendor to achieve a higher score on the Supplier Rating Systems in place. This survey has focused on ranking 39 proposed selection criteria in order of importance from Buyer's perspective. Further work could be carried out to determine success rate as a performance indicator of suppliers selected using the said process.

The study focused on chemicals manufacturing companies in India and the process and attributes they consider for selection of suppliers. Since the approach used in the study was based on standard selection criteria, it is intended to cover issues such as quality, capacity in terms of finance, services, responsiveness, etc., whereas, chemical industries are in large connected with a lot of environmental issues especially in high growth regions like China, India, Brazil, etc. therefore, exploring for "Green Supplier Selection

Criteria" for an organisation to respond to any existing global trends w.r.t. environmental issues related to business management and processes would be important for a sustainable solution.

The business environment has increasingly become dynamic, and the interest of members in the business process including governments, customers, suppliers and competitors have been changing in response to competition, technological changes and public concerns. The tendency of an organisation to integrate their supply-chain in an attempt to reduce costs and be able to serve their customers better and changes in environmental requirements and public pressure on organisations have become a new trend in the business world. These trends together provide an opportunity to exceed environmental expectations of government and customers through supply-chain collaboration; these include an ability to see supply-chain member's commitment to environmentally friendly practices.

There has been an increasing concern for sustainable economic development all over the world. Governments are trying to adjust legislation and social pressure through individual activists, non-governmental organisations, and the international institution is also growing to express public mandate against the negative impact of business activities on the environment. Internally, the need to improve organisational efficiency, reduce waste, overcome supply-chain risk and achieve competitive position has made companies start considering environmental issues from a competitive viewpoint.

The level of inter-organisational relationship or collaboration determines the extent to which an organisation can improve its environmental management performance and therefore become either a reactive or proactive organisation. The reactive organisations focus mostly on their internal functions and their responses vary from resistant adaptation (to avoid penalties and public pressure) through receptive adjustments of the current processes, to constructive responses which are more efficiency based but with secondary environmental advantages, some of these examples are:

- Reduction of pollutants emitted into the air instead of pollutants produced.

- The inclusion of basic environmental clauses into purchasing contracts to seek compliance with regulations.

- Use of established international environmental standards like ISO 14001.

Whereas proactive strategies involve the Closed Loop Supply-Chain, Total Quality Environment Management in Planning and Operations, and Product Life Cycle Cost Analysis. TQEM is applied for both products (specific design, formulation, characteristics and functionality) and for processes (production, distribution and use).

The purchasing function in relation with other functions has a greater role to play in environmental management performance of an organisation. Having seen and explored the advantages of right supplier selection process for sustainable business gain and win-win relationship with customer and supplier and making it obvious on which supplier to collaborate and how to select suppliers for the organisation's performance. Incorporation of objective environmental criteria in the evaluation system for the chemical industry will ensure higher environmental performance in the collaborative supply-chains.

Green supplier selection criteria may be developed with the intent of focusing on meeting government regulations, focusing on process improvement and focusing on buying the company's environmental policy and these green criteria can be classified into Qualitative (Green Image, Environment Management System, Environment competencies, Design for Environment) and Quantitative (Pollutant costs/effects, improvement costs) criteria.*

Challenges of Today's Sourcing Manager

Purchasing often fails to get management's respect because its costs saving reports aren't believable. But you can make your costs saving reports more believable if you can understand and reconcile the difference between cost savings and expense reductions.

When you claim cost savings, management expects to see the expenses on its financial statements lower than the previous year. In many cases, the cost savings you report are much different than the change in expenses.

This disparity hurts your credibility. Managers with a financial background were trained early in their careers to reconcile financial reports; ensuring that differences between numbers from two sources are accounted for.

You should account for the differences between your cost savings and the actual expense reduction. Here are reasons cost savings exceed expense reductions. Think about how to reconcile these in a cost savings report.

Quantities Increased: If you paid $2.00 per pair of earplugs last year and this year purchased 10,000 earplugs at $1.50 per pair, you'd probably report a cost savings of $5,000 ($0.50 × 10,000), right? But what happens to the expenses if you only bought 5,000 last year? That's right, the expenses increased from $10,000 (i.e., $2.00 × 5,000) to $15,000 ($1.50 × 10,000)! So, account for that when reporting cost savings, explaining that while the quantity increase would have resulted in expenses rising by $10,000 if last year's price was held, you offset $5,000 of that increase by improving pricing.

New Expenses Arose: If you purchased large quantities of an item that your organisation never used before, you don't have a price benchmark.

If you negotiated the low bidder's prices even lower, you'd likely report the negotiated difference as your cost savings. But what do expenses do? They go up because you spent money on a category on which you spent no money last year.

Increases Aren't Reported: Cost savings are often reported in isolation. Let's say you reported cost savings of $100,000 for reducing costs in one category by that amount. But what if the spend in your other categories increased by a total of $200,000? Should you still report $100,000 as your annual cost savings?

Better-Than-Market Performance Is Counted: Many purchasers measure their cost savings on price-volatile commodities (e.g., petroleum-based products) by comparing percentage cost changes with indexed cost changes in the market. But before claiming cost savings when you beat the market, consider how management may see things in terms of expense reduction. One of the most important practices in demonstrating Purchasing's value to the organisation is tracking and reporting cost savings. When you achieve cost savings, there is important data you should log, such as:

- Date

- Description of Product or Service

- Baseline Price

- Type of Baseline or Type of Savings

- Current Price

- Quantity

- Cost Savings

- Buyer

- Supplier

- Cost Centre

- Category or Commodity

Baseline Price. Costs saving is the difference between your current price and a higher price called a baseline.

Type of Baseline or Type of Savings: Baselines may include last year's price, originally proposed price, the cost to produce internally, etc. The baseline may determine whether management deems the impact of an avoidance or cost savings and may only be interested in the latter.

Cost Savings: This field should be automatically calculated by multiplying the quantity by the difference between the baseline price and the current price.

Cost Centre: Reporting your cost savings by cost centre helps Purchasing demonstrate the positive impact it made on a specific department's budget.

What Are The 4 C's of Cost Containment?

If you're looking at the letters in the term "Cost Containment", you'll only see two "c's". But there are actually four "c's" associated with the practice of cost containment. Use these four techniques to maximise the results from your cost containment initiative.

Consolidation – A timeless principle of cost-effective purchasing and supply management is that the more you buy, the less you pay per unit. So, you need to consolidate your enterprise-wide spend in a category or combination of categories into a large "Market Basket" that you will entice suppliers to bid on. Leveraging your volume in this way helps to maximise the cost savings available to you.

Competition – Competition between suppliers is a powerful tool for reducing costs. After you have developed your market basket, you should identify a healthy number of suppliers to bid on your requirements, perhaps even using a reverse auction, if appropriate. The less certainty that suppliers have about earning a valuable chunk of business, the more aggressive their pricing will be.

Contracting – What gives suppliers a lot of incentive to win your business? A long-term guarantee of a certain volume of business does! If you are able to promise a forecasted quantity to the successful supplier via a contract lasting three or more years, you will likely inspire in suppliers a keen interest in helping you achieve your cost containment goals.

Collaboration – Your cost containment work doesn't have to end when the contract is signed. You can collaborate with your supplier to identify waste, inefficiencies and other opportunities to take cost out of the supply-chain in a manner such that both your organisation and the supplier can financially benefit. Agree to meet regularly during the term of the contract with the goal to exchange ideas that can result in measurable cost containment activities.

With a large market basket, maximised supplier competition, a contractual guarantee of multi-year volume and a plan for working with your suppliers, you are well on your way to cost containment success.

HOW CAN SPEND ANALYSIS CONTRIBUTE TO SAVINGS?

Simplified, Spend Analysis is the systematic review of historical purchase data. The output of a spend analysis is a summary of purchases by various variables such as category, supplier and/or business unit.

The primary reason for conducting spend analysis is to identify opportunities for cost savings. When you look at the output of a spend analysis there are at least four indicators that there are opportunities for cost savings:

Indicator #1: A large amount of spend in categories for which enterprise-wide contracts do not exist. However, when these best practices have not been utilised in a category or purchase decisions are left up to non-purchasing professionals, there is a likelihood that the company is overpaying for goods and services. So, identifying areas where Purchasing should be more involved in getting contracts for those categories is a sound strategy for cost savings.

Indicator #2: A significant Purchase Price Variance (PPV) for a high-spend item or category. PPV is the difference between the average price paid and a standard cost. When there is a large PPV, this indicates one of two things: either your standard cost is not valid, or you are paying too much. In the latter case, you should consider taking some type of action, such as negotiating or sourcing, to ensure that you are paying a fair price.

Indicator #3: An unusually large number of suppliers for the money spent in that category Purchasing 101: the more you buy from a single supplier, the better discount you'll qualify for. If you're buying from too many suppliers, you're not leveraging your volume and likely not maximising your discounts. Seeing a large number of suppliers in a category can tip you off that supplier consolidation can deliver savings to your organisation through price reduction and also through the more advanced benefits of an optimised supply base.

Indicator #4: Rising prices over time. If you're paying more year after year, you simply need to pay more attention to the purchases in that category. With no one minding the store, price creep is likely to set in.

Cost savings is the most common performance metric executives use to evaluate their purchasing departments. So, it's critical to have strong cost savings report.

Let's examine three problems found in weak cost savings report.

Problem #1: Using an Incorrect Baseline. Cost savings are calculated as the quantity to be purchased multiplied by the difference between the price you will pay and some higher baseline price. Make sure that your company's executives agree with the baseline you use. Picking a high but not credible baseline (e.g., the highest bid received) may maximise your cost savings calculation but hurt your credibility with those executives.

Problem #2: Using Poor Quantity Estimates. Because the price you will pay is lower than the baseline price, your cost savings total grows as you purchase more. You may report a certain cost savings value based on quantity estimates. But what if you only end up buying half of that quantity? That's right – your actual cost savings will be half of your estimated cost savings. When you estimate your cost savings to executives, they expect the company to realise the cost savings you estimate. If the cost savings realised is less than your estimate, you won't be their favourite employee.

Problem #3: Failing to Adjust Budgets. To executives, "Cost Savings" is synonymous with "Profit Improvement". When you say that you will save $500,000 this year, they expect profits to be that much higher. So, let's say that you achieve cost savings of $500,000 for a department within the company, where does that $500,000 go? Does it free up money that they can spend on other things that they hadn't budgeted for? If so, then your cost savings is not really a profit improvement. The company is still incurring the same amount of expenses. To truly improve profit, an amount equal to cost savings must be removed from the budget. Purchasing obviously can't do this alone but working with senior management in this respect can certainly help executives see the value of intelligent purchasing.

BALANCING IS KEY TO SUSTAINABLE SUPPLIER PERFORMANCE

The primary objective is to identify criteria which can be effectively used for screening of suppliers before the start of real business. During this research project, an effort was made to identify various important attributes and based on the survey results and respective weight to various attributes and prioritisation, we have tried to develop/frame screening template.

There is a flexible provision for weight score, based on the individual company or industry-specific dynamics. The chemical industry in India or world goes through various types of controls and regulations like REACh, GMP, etc. and these rules differ from country to country, i.e., based on supplier country as well as to accommodate regulations of customer country + customer company-specific requirements.

Based on the research survey, each main category got defined into weighted average score on 1 to 10 scale. For instance, there is a weight of 10 for contamination and prevention attribute, which is very important criterion in Food, Pharma, Agro and Electronics chemical industry and might not be as important for textile or construction chemicals. Thereafter, suppliers are rated on 1 to 5 scale on these attributes, 1 mean low and 5 means high. Both these scores get multiplied to reach the final score for each attribute and then the score of each attribute is added for a total score. In order to set-up minimum qualification, it is recommended to make 1000 as minimum qualification score in addition to the following given category-based score.

For instance, if a company XYZ is conducting supplier assessment test they can decide weight between 1 to 10 for each attribute and then the audit team of XYZ can rate supplier organisation between 1 to 5 for various attribute and later as a team agree to one single score between 1 to 5 for each attribute and multiply that score with weight decided between 1 to 10 to reach to final score. This can be repeated for all the 39 attributes, and if the total final score is over 1000, the supplier can be qualified for a further business discussion or trial supplies or development work, etc.

Low-cost country sourcing involves a more holistic business perspective when evaluating and selecting suppliers. First, a long-range assessment of the supplier's overall business health and process capability, then understanding supplier strategy so that new areas can be discovered for a strategic relationship which can add value and suppliers become a more integral part of the company operations. It is critical to foresee any conflict of interest the supplier may encounter to derail the relationship while conducting this detail supplier assessment.

Illustration w.r.t. Qualification Screening of Chemical Supplier

Company Goodwill/ Market Reputation	Min. Total Score	Attribute Weightage	Supplier's Ranking	Total Score
Proprietary Information and Protection		10	5	50
Company Vision Alignment		7	3	21
Not a Generic Competitor		8	1	8
Historical Relationship		8	4	32
Value Seen in Association		9	3	27
Communication System		9	5	45
People and Organisation	160	Avg. 8.5		183
Labour relations record		7	3	21
Training and Orientation		5	4	20
Willingness to Change/Flexibility of Organisation		10	4	40
Organisation Efficiency and Response		10	5	50
Quality of Human Resources		7	5	35
Leadership Commitment		8	5	40
Delivery and Logistics	150	Avg. 7.90		206
Reliability to deliver and perform		10	5	50
Lead time		10	5	50
Plant Location		7	3	21
Warehousing to deliver on time to request		9	4	36
Product Availability		10	4	40
Procedural Compliance		9	4	36
Supplier Quality Management	170	Avg. 9.20		233
Product quality and consistency		10	5	50
R&D and Analytical capability		9	4	36
Product Performance		10	5	50
Warranties and Claim Policy		9	4	36
Packaging ability		9	5	45
ISO and Other quality certifications		9	4	36
Corporate HS+E and Social Responsibility	170	Avg. 9.40		253

Company Goodwill/ Market Reputation	Min. Total Score	Attribute Weightage	Supplier's Ranking	Total Score
Compliance of law and Ethical Standards		10	4	40
Health, Safety and Environment compliance		10	4	40
Signatory of responsible care programme		5	4	20
Documentation compliance and knowledge		8	4	32
Exposure to related regulations		8	4	32
Product registration and handling		6	4	24
Financial Soundness	**120**	**Avg. 7.80**		**188**
Financial soundness to make investment		7	4	28
Global cost competitiveness and sustainability		10	4	40
INCO/Pay terms flexibility		8	4	32
Manufacturing and Innovations	**80**	**Avg. 8.3**		**100**
Manufacturing capacity/capability		9	3	27
Contamination and Prevention		10	5	50
Continuous improvement programme		7	4	28
Process and Facility standard		8	5	40
Technical Expertise for NPD		6	4	24
Technology and Engineering Support		8	4	32
	150	Avg. 8.0		201
	1000			1364

Here above after considering the importance and weight to various attributes (based on survey analysis) given by respondents, I've tried to develop a "Supplier Selection Screening" template as an outcome of this study report, which addresses minimum score requirement for overall qualification i.e. 1000 marks, as well as minimum marks required by any organisation to get qualified for business initiation on category/Sub-Categories basis w.r.t. Company Goodwill, People, Delivery, Quality, HS+E, Financial and Operations. Above illustration has taken average score from survey analysis for various categories but based on the business/ company-specific needs re-distributed weightage for sub-categories, giving flexibility to the organisation to tailor-shape this template to suit their specific requirements or adjust to local market dynamics or industry conditions to meet their customer expectations. Further, based on all above analysis conducted with respect to Management's, Buyer's and Customer's perspective, the following attributes have surfaced as top ten attributes where the company should focus on before selecting a potential supplier for a business relationship;

Ranking	Top Ten Attributes for New Supplier Selection in Chemical Industry
01	Product Quality and Consistent Performance
02	On-time Delivery as per requirement of Customer
03	Contamination and Prevention
04	Confidentiality and IP protection
05	Global Cost Competitiveness
06	Compliance of Law and Ethical Standards
07	Manufacturing Technology and Innovation
08	Company Goodwill/Market Reputation
09	Corp. HS+E and Social Responsibility
10	Organisation Efficiency and Response

On the other hand, during the course of this study based on discussion and interaction with various small, medium and large-scale suppliers, the following key best practices emerged as major success factors in sourcing projects;

- **Become Business Partners, Not Just Buyers** – overcome a pervasive buyer mentality. Buyers tend to seek the lowest possible price regardless of any other benefit. Procurement should identify and respond to broader business goals proactively. Partnering with low-cost suppliers involves a mindset shift–from price to value, from products to the solution, from input to output.

- **Explore New Value Frontiers; It Is Not Just About Price** – Focus should turn to an overall business outcome, the total cost of ownership and the potential for creating long-term value for the company.

- **Make Suppliers Part of a Team; the Best Value Chain Wins** – Must take a more inclusive approach on supplier's management, should bring them into the decision-making processes and change initiatives.

- **Low-Cost Country Sourcing Can Deliver Tangible Results to the Bottom Line** – Working with a 3rd party may open the door to minimal upfront capital investment and combined with a holistic approach to procurement transformation, be a quick route to such savings.

- **Know the Suppliers As Well As Yourself; Smooth Volatility with Predictive Demand** – Predict demand and be able to react to demand variability with rapid response and allocation of resources to meet organisation expectations.

- **See What Others Do Not; Unveil Visibility with Collaborative Insight** – Collaborate with visibility to events with suppliers, service providers and customers in an open, action-oriented environment.

- **Exploit Supplier Efficiencies; Enhance Value with Dynamic Optimisation** – Create cost-efficient, sustainable products and services while hedging risks with partners.

Secondly, the majority of companies agreed that supplier selection criteria for a product or service category should be defined by a cross-functional team of representatives from different sectors of the organisation. Team members should include personnel with technical/applications knowledge of the product or service to be purchased as well as members of the department that uses the purchased items to ensure that selected supplier has all the required competencies for a sustainable relationship. This interaction will also be helpful for the supplier in understanding the product and organisation requirements and expectations w.r.t. quality, process, documentation, batch size, timing, stocking, etc.

The following were the additional comments/criteria shared by respondents:

- Previous experience, i.e., years in business and past performance-related records of the supplier, including supplier's track record for business performance improvement.

- The relative level of sophistication of the quality systems, including meeting regulatory requirements or mandated quality system.

- Ability to meet current and potential capacity requirements and the ability to do so on the desired delivery schedule as well as customer service.

- Technical support availability and willingness to participate as a partner in developing and optimising process improvement and a long-term relationship.

- The total cost of dealing with the supplier (including material cost, communication methods, inventory requirements and incoming verification/testing requirement and cost).

One conclusion as a supplier one must focus is to get to know the real need of their customer base rather than presuming their highly visible wants and making their entire Sales and Marketing Strategy in accordance.

ACCEPT CHANGE AND BE A CHANGE LEADER

BEST PRACTICE IN PURCHASING?

The answer is–"NO".

Practices are created every day according to the situation. There is nothing like standard purchasing practice. It is all about attitude and strategy to meet different situations and challenges.

You don't go to a library to pick up a book on Purchase Management to address a particular situation considering the speed of response desired and uniqueness of the circumstances faced by you which are never a part of any readily available guidelines.

Therefore, the BEST PRACTICE is to accept constant change, allow your mind to think through the appropriate solution which can ideally meet the requirements of the situation and share the knowledge with people around you to become a change leader. In today's dynamic business scenario, it is mandatory to be innovative, informative and transparent in everything that we do. Risk-taking ability, creating benchmarks for others and setting standards that others can match up to, are thebest practices.

The purchasing arena has undergone a lot of changes and to meet the challenges in the constantly changing landscape, a few areas which I believe are of paramount importance for any purchasing activities are listed below.

Transformed Decision-Making

In the ever-changing business scenarios, the ability to take quick, appropriate decisions is the necessity to capitalise on the right opportunity. This means empowerment and strong teamwork.

This has brought about a major change in the decision-making process and entities. Today, decisions are driven by professionals instead of owners. To arrive at a decision, the professionals involve concerned cross-functional team members to ensure compliance and appropriate match to the requirements. This shift has brought about transformation in the role of procurement department as it is no longer simply buy goods/services as desired for manufacturing and clients but has become a vital link in the business change over process.

Essentials of Effective Purchasing

Purchasing functions are carried out to ensure a consistent supply of quality material in time at the lowest cost to meet production or client requirements. In order to carry out this function most effectively, it is desired to,

- have coordinated and cost-based procurement
- have cross-functional team driving procurement decisions
- have strategies, vendor partnership and vendor economic driving the procurement decision
- have globally competitive buying, clubbing all the above together effectively

Highly Informative Team

Information is the key tool which enables purchasing professionals to negotiate most effectively. Today, both the buyer and the seller teams are highly informative and come to the negotiation table not only with a lot of knowledge on product and market but also with an understanding of the manufacturing process as well. Both consider global sourcing as a major parameter in striking a deal. A lot of transparency is exhibited, and the information is shared without any hesitation.

Analysis of Total Spent

One of the best practices proved to be time-tested and successful is the analysis of total spend for your organisation.

- Do a one-year analysis of the total spend of your company on raw material
- Break them into major finished products in the order of highest contribution in profit and volume.

- Identify the impact of key material in those high impact finished goods.

- Set your goal in becoming highly cost-efficient in this material.

- Work out the cost-plus formula for this material by completely understanding the chemistry of the product, its usage norms, wastages, recoveries and the cost thereof, utility cost, labour charges and ultimate processing cost.

- This would give you a firm idea about the product you are going to buy leading to a very healthy negotiation and highly competitive prices.

Assume your total spent analysis look like the following

PRODUCTS DISTRIBUTION OF TOTAL SPENT OF 1000 cr.

PRODUCT D,	40	25%
PRODUCT E,	10	6%
PROUDCT F,	5	3%
PRODUCT G,	35	21%
PRODUCT C,	25	15%
PRODUCT B,	20	12%
PRODUCT A,	30	18%

- Identify combination of products which contributes maximum sales revenue for the organisation.

- Track the top few products which are the highest contributors on this list.

- Again, identify the KRM (Key Raw Materials) being purchased from the manufacturer of this product.

- Study this raw material in detail and identify the key input in the manufacturing of same.

- Repeat the steps you took in identifying your key raw material assuming that you are the buyer of the key input being purchased by your vendor.

- Such an in-depth analysis of your vendor's raw material would throw open several options for you.

- Combined buying of key input maximising quantity discount and best possible rates and conditions.

- Entering into rate contract with the vendor on Job Work or Conversion basis.

- Suggesting different manufacturing process which may result in better yield and lower utility cost.

- Changing the consumption norm of the input going into the manufacturing of the raw material.

- Usage of alternative input, i.e., lesser purity material which is available at a much cheaper rate than proportionate to the purity and the end cost would be far cheaper than the manufacturer using higher purity material.

- Set up your own manufacturing unit for this key raw material and open up another avenue of sale after meeting your captive requirement.

- Buy out the facility exclusively for your own use.

- Guarantee a fixed margin to the manufacturer and get a competitive rate for you and bind the manufacturer to sell the product to all other at a rate higher than to you.

In conclusion, the best practice in purchasing is determined on your adaptability of change, having desired information, carrying out spend analysis to decide your focus and most importantly involving your vendors as your business partners, to ensure procurement of desired goods in time at minimum cost.

Therefore, to name a few selected best practices across the industry in sourcing function are as follows;

- The total cost of ownership

- Portfolio Analysis and Zero-based management

- Cost Break down analysis

- Competitive sourcing

- Risk Analysis

- Reverse Auction

Internal Barriers and Roadblocks

Whenever people work together, they are going to disagree about some things. Underlying disagreement will be the entire fabric of their history together; their personalities and the urgency of the issue can create a pressure cooker. Handling disagreement at work can be tough, and at times, it creates roadblock; it may require dealing with egos and occasionally with childish behaviour. Try to keep everyone happy or change into a middle determined way to win.

Procurement teams today must focus more than reducing costs to build supply networks that contribute to value and profitability. The challenge lies in fundamentally changing how organisations and procurement teams especially, viewed within the organisation. If senior leadership decide what they think is best then at times it's quite challenging for procurement team to win alignment of cross-function or even convince senior leadership for investing time and money for some futuristic aspect like qualification of new source when everything is going fine with the current suppliers, especially if there is no recorded issue with the supplier. Probably we will never get R&D and business alignment to drive such projects as it does not sit in their priority and everyone is so busy that they can see a problem with the sole-source situation. In such a scenario, if procurement doesn't address such issues proactively, when a business has time and can do comprehensive way alternate supplier development, then in crisis business has to do it later.

In many organisations, there is always an unwritten organisation chart or structure that plays advisory role not as a solution provider but more as a barrier, which eventually puts Business, Technology, Engineering and Finance, etc. on top of pyramid and procurement is viewed as a lower and more transactional function in place of strategic wing. Many buyers are measured tactically (e.g., by the number of events they run). This tactical view of procurement sends a strong message to those who run the business that there's not a lot of criticality to what they do within the context of business goals, reducing motivation to do more. Often accompanying this roadblock is that the leadership team doesn't fully understand the contributions procurement team makes to the company's overall performance and won't back investment in procurement to improve the situation.

Really good procurement professionals know their markets, the best and the worst players in the market and how to work their contacts. They've spent years nurturing and growing relationships, believing that's the best and only way to get what's needed from suppliers–lowest cost, improved service levels or better terms.

Today's supply networks are complex. Supply-chains made up of hundreds or thousands of suppliers, many around the world, are not unheard of. Category requirements are so complex that it takes an experienced procurement professional to understand the complexities of defining what's needed, evaluating what's offered and negotiating the pitfalls. In many cases, these products and services are strategic to the company. This roadblock often arises at the same time as over promise and under deliver and is accompanied by the question, "Are you sure you want an outside sourcing team or technology application to handle something that critical?" Regardless of where your organisation is in its sourcing maturity, there's room to make a bigger impact. There will always be hurdles when trying to transform and improve. As we stated earlier, change is hard. But by identifying the root causes and putting a plan in place for corrective action.

- *Over Promise and Under Delivery Won't Be Very Healthy Sign for the Business* – Organisation or procurement function need to come out from this syndrome, at times this mindset gets generated due to feeling like, "I know my supplier is giving the best services". This can't be true unless we create competition and benchmarking.

- *Bureaucratic Structure Reduces Risk-Taking Ability of the Team* – Leadership may not see inefficiencies and pain in the same light as those who are experiencing them. Keep a bigger picture approach w.r.t tangible benefits to the company and implementation time frame, and a reason to take action now.

- *Technology Non-Suitability Can Create a Barrier to Progress* – Selecting collaboration software that does not gel with the IT resources, budget and integration requirements of both buyers and vendors can be the single biggest hurdle to bilateral collaborationon. Technology that is flexible enough to support a wide array of external ERPs, internal systems for planning, procurement and finance, along with disparate connectivity protocols, can not only significantly reduce barrier but actually enable more suppliers, partners and internal groups to collaborate via the same platform.

- ***Short-Term Focus Can Derail Long-Term Objective of Procurement Function*** – Collaborative efforts between companies may not show significant returns. If management is incentivised based on short-term results, then investments in long-term stability can be a hard sell. At times there is a strong feeling that procurement is tactical and there is no need to tie it to business goals, can prove a big hindrance.

- ***Win Or Lose Mindset Will Create Unsustainable Process*** – Unless one side has enough influence to simply impose a solution on the other side. More advanced forms of bilateral collaboration include continuous replenishment, Vendor Managed Inventory (VMI) and collaborative planning forecasting and replenishment. If you are trying to convince a vendor to join your collaboration initiative, make sure they see a tangible benefit to continue doing business with the company. Only one company benefit will never be a sustainable equation.

Therefore, if the above barriers are not removed on timely phase manner, it may become roadblock for this function to grow and create a broader outlook for the organisation.

SOCIALLY RESPONSIBLE PURCHASING

Socially responsible purchasing is a hot trend these days, but there's one topic that's been missing from every set of socially responsible purchasing guideline, i.e., Animal Welfare. This aspect of socially responsible purchasing will escalate soon for many reasons. One is that the media has become very interested in exposing animal cruelty scandals.

The increased exposure to animal cruelty is resulting in pressure on executives to eliminate the atrocities that corporate spend supports. That pressure will prompt edicts to adopt animal-friendly purchasing practices.

There are many spend categories that potentially involve animals. Three present in nearly every supply-chain are leather furniture, leather laptop bags and cleaning supplies that involve animal testing. "Many people think that leather is a byproduct of the meat industry and that is simply not true." Animals used for leather suffer gruelling transportation during which they are often deprived of food and water, and their handlers will break their tails and rub cayenne pepper in their eyes to get them going.

Upon reaching the slaughterhouse, the animals' throats are slit, and they are often skinned alive. For some cleaning supplies, animals are unnecessarily forced to swallow or inhale massive quantities of a test substance or endure the pain of having their eyes or skin chemically burned. Now many companies that have adopted animal-friendly purchasing practices including Whole Foods who include quality of animal welfare programmes as a point in their supplier rating system.

There are four simple steps for launching an animal-friendly purchasing programme:

- Survey suppliers about their animal testing practices and how they monitor their suppliers' animal testing.

- Start with a specific area in which to adopt a no animal testing and/ or no-fur policy.

- Give purchasing preference to suppliers that provide synthetic alternatives and no animal testing.

- Conduct unannounced audits to determine if your suppliers are engaged in animal-friendly practices.

Tactics for Managing Supplier Relationships

Supplier Performance Evaluation – Ask a supplier's representative how they think the supplier is performing and you may hear "Great!" But what if you think the supplier is performing poorly? Who is right? You can't tell without agreed-upon performance standards. For your strategic suppliers, agree upon what to measure (e.g., the percentage of orders delivered by their due date) and what the goal is (e.g., 95% on-time deliveries).

Idea Sourcing and Value Creation – Better profitability can come from ideas. You can greatly increase the number of good ideas by sourcing ideas from your suppliers rather than just from your company's employees. Some leading organisations have systematic processes in place to collect ideas from suppliers, measure their impact and reward suppliers for them.

Supplier Development – It's logical that when you improve the capabilities of your company's workforce, your company's benefits. But even though suppliers now do the work which is once done in-house, that logic hasn't followed the work. Leading companies engage in supplier development, providing resources to improve their suppliers' capabilities. This often involves training suppliers in methodologies such as Six Sigma or Lean, but really can be any collaboration that makes suppliers more capable of delivering benefit to your company.

A Joint Review of Purchase Costs – If you work for a big company, you have a lot of buying power. Buying power that may be wasted if your smaller suppliers have 100% responsibility for buying all of the goods and services needed to provide their product or service to you. By jointly reviewing costs further down the supply-chain, you may find opportunities where you can buy some goods and services your suppliers need at a lower cost, ultimately reducing your overall costs.

SOURCING SKILLS AND PROFESSIONAL DEVELOPMENT

What is the difference between Purchasing and Supply-Chain Management?

Though many different and conflicting definitions of Supply-Chain Management are available, Purchasing is a subset of Supply-Chain Management. Purchasing deals primarily with managing all aspects related to the inputs to an organisation (i.e., purchased goods, materials and services), while Supply-Chain Management deals with inputs, conversion and outputs.

A supply-chain consists of three types of entities: Customers, Producer and the Producer's Suppliers. The extended supply-chain includes customers' customers and suppliers' suppliers.

Supply-Chain Management oversees and optimises the processes of acquiring inputs from suppliers (purchasing), converting those inputs into a finished product (production), and delivering those products or outputs to customers (fulfilment).

Under this definition, supply-chain managers decide where to locate manufacturing and distribution facilities, how to route goods and materials among those facilities, and from which parts of the world to source the inputs. Supply-Chain Management unites disparate functions that historically reported to different executive positions with different, and sometimes conflicting, priorities.

So, what does this mean for individuals who have a purchasing-related title?

One myth is that purchasing will become less important. To the contrary, analysing spends on cost savings opportunities, negotiating and selecting reliable sources of supply will always be critical. These functions fuel profit and provide a competitive advantage for the organisation.

However, the purchasing professional can expect to see his or her role expand to include the management of functions that were separate in the past. These functions include inventory management, internal logistics, warehousing and other functions that are more related to the input or pre-production side of the supply-chain. Today, due to this expanded role, Purchasing is often referred to as "Purchasing and Supply Management". Therefore, individual working in this segment needs to upgrade their skill set on a regular basis whether related to sourcing or supply-chain related activities.

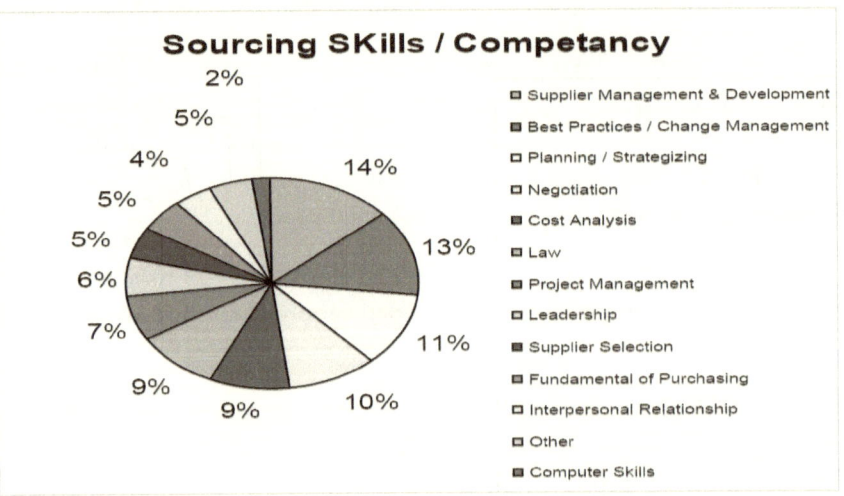

Great leaders drive great performance. To attract, develop and retain the highest calibre of sourcing talent, we need to: articulate specific behaviours which business and operations leaders can expect from sourcing leaders; describe what it takes to move into positions of greater responsibility and scope, and define clear development actions that help people achieve success throughout their career.

The Behaviour Matrix provides sourcing professionals with a guideline of how the behaviours apply specifically to expectations for sourcing professionals wherever they are in their career.

1. Customer Focus

2. Work Planning and Implementation

3. Strategic Thinking Style

4. Knowledge of the business environment

5. Negotiation skills

6. Personal integrity

7. High level of purchasing expertise

8. Performance orientation and Self-motivation

9. Measurement and Forecasting

The term "customer focus" is very wide; we say often that anything we are doing and not adding value to our customer is actually a waste, and our customer should not pay for that waste. While selecting an individual for sourcing job, it's important to measure individual ability to anticipate and meet customer needs at the appropriate level and time and he or she should have the drive to continuously strive to increase customer satisfaction.

Full implementation of the purchasing category and supplier management process across all expenditure calls for an ability to drive projects and process change according to demanding timetables. Work planning also focuses on supplier management long term and in a standard way and needs to be driven at every opportunity of contract with the supplier.

The incumbent should have strategic thinking style and would require a sound knowledge of selected supply market relevant to business categories of expenditures. He or she can describe the major features of such markets and knows where and how to obtain relevant information.

We have already discussed the the importance of negotiation skills for sourcing professional. This calls for the expertise in both tactical and strategic negotiation.

Personal integrity is an important attribute, the incumbent should be able to work with confidential information and must demonstrate principled leadership and sound business ethics, show consistency among principles, values and behaviour and build trust with others through own authenticity and follow through on commitments.

High level of purchasing expertise is equally important as the person is going to be responsible for defining and implementing purchasing category best practice, strengthening supplier management and driving supplier reduction/addition forward.

The ability for performance orientation and self-motivation for driving result is important; he or she should be a self-starter who delivers projects on time. Notably strong in self-management demonstrates superior skills in developing plans and managing programme execution across functions and sites. Drives project activities without the need for regular upward reference to higher purchasing levels.

In order to elevate functional excellence in procurement to new highs, an organisation should devote itself to the following areas:

1. Risk vs. Spend analysis

2. Segmentation of needs

3. Analysis of supplier's market

4. Portfolio analysis of suppliers

5. Prospecting suppliers and selecting suppliers to portfolio

6. Define and float RFQs

7. Negotiation and establishment of contracts

8. Evaluate, monitor and optimise supplier performance

9. Make continuous improvements

10. Using e-procurement tools

It all starts with people. The entry barrier into a procurement career must itself be quite challenging so that only the best professionals make it. Continuous focus on training and developing people and the acquisition of new skills must be accorded top priority. An encouraging and challenging work environment must be created at the workplace where people bring in new ideas (innovation), are adequately compensated and look forward to a great career ahead.

Sourcing Tips

1. **Be Environmental Friendly:** Incorporate environmentally responsible practices in your supply strategies, such as the use of recycled material, sustainable energy. These efforts ensure compliance with tougher environmental regulations and mitigate price increases.

2. **Partner with Logistics:** Transport handling and tariff cost range from 13% to 24% of the basic price of imported material, see how to lower the total landed cost by using various tariff codes.

3. **Cultivate Safety Culture:** Incorporate safety practice in supplier development, such as near miss, unsafe behaviour, etc. It's cost-effective in the longer run.

4. **Share Risks with Supplier:** Ensure that your contracts include clauses to have supplier share risks in areas of payment terms, currency fluctuation, capital investment and freight surcharges: this will reduce exposure and create a true partnership.

5. **Get Supplier on Design Team:** Secure a spot for your most strategic supplier early in your product design phases. It ensures that the supplier has the capability and capacity to support your new design. It has been proven to reduce costs and speed development cycles.

6. **Be the Middleman:** Leverage your volume to buy commodities/parts at a lower price and resell to supplier partner to reduce total products/supply-chain costs. This can lower your supply cost and if managed properly, can let you turn a profit on the resale.

7. **Work on the Concept of Co-Operative:** Aggregate spending or share contract. Some suppliers such as contract manufacturers often secure better deals than you on certain parts and components.

8. **Educate Your Supplier Well:** Share your sourcing and supply management expertise and system with your Tier 1 suppliers. This approach can increase your visibility to sub-tier supply costs and risks.

9. **Find Hidden Experts:** Get a crash course in new spend categories from functional stakeholders who understand cost drivers and innovation in those areas. For instance, Tap IT for software and hardware, you bring the negotiation knowledge and the lunch. You'll find new cost-saving opportunities.

10. **Finance Your Future:** Have finance validate supply saving and track policy compliance.

11. **Get on the M&A Squad:** Nearly 60% of the hard dollar impact of any M&A deal comes from Supply Management improvements. Let your CEO know the impact Supply Management can contribute to the deal.

12. **Share Gain with Supplier:** If a supplier provides a cost reduction or improvement suggestion that is successfully implemented, share a portion that is successfully implemented and share a portion of the saving with the supplier. This will encourage suppliers to share proactively, contribute ideas that eliminate costs and provide a joint competitive advantage.

13. **Harvest Supplier Innovation:** Establish formal channels and procedure to receive, evaluate, implement and measure the results of cost reduction or improvement suggestions from suppliers.

14. **Solicit Feedback:** Top performers use a 360-degree performance review, actively soliciting supplier opinions on how they rate as the customer will improve on supplier relationship management.

15. **Give Frequent Report Cards:** Establish standard, easy to understand metrics for measuring supplier performance. Share performance scorecards with supplier frequently, giving them ample chance to refute and improve their score.

16. **Create Your Own Opportunity:** Volunteer to take charge of your company supply initiatives. If you lack skills educate yourself on supply management technology.

17. **Check on Supplier's Well-Being:** Keep periodic checks on the financial and business health of your suppliers. Pay attention to any liens, legal judgements or restructurings.

18. **Look Inward Before Outsourcing:** Document and standardise processes before outsourcing spend categories. This helps define expectations, gain alignment and mitigate risks and failure later.

19. **Go LCR:** Invite qualified suppliers from low-cost region to participate in negotiations. This may drive other suppliers to bid more aggressively.

20. **Ask "Why Not?":** Why can't this spend category be e-sourced? This discipline will ensure consistency. Documents sourcing method are used for all spends and provide vital insight into spending and market trends.

21. **Sell to Suppliers:** When rolling out new initiatives or technologies, educate suppliers on the benefits they can expect and how best to use your tool of choice. Better trained suppliers make more competitive offers.

22. **Buy Lunch:** Host periodic lunch and learn workshops to train team members and other functional stakeholders on your Supply Management initiatives.

23. **Analyse Before Auctioning:** To determine which spend categories are ripe for auctioning, ask four questions. Can you define clear specifications for the category? Is there a competitive supplier pool? Is spending significant enough to be of interest to suppliers? Are you prepared to change suppliers based on the auction results? If yes, then auction it.

24. **Go Old School:** Gain insight into supply market pricing and availability trends for specific categories by using traditional market intelligence methods. Tap your personal contacts for their own experiences and supplier recommendations.

25. **Listen to Your User:** Solicit stakeholder's feedback on process improvements or solution functionality changes. This secures stakeholder alignment and improves adoption and compliances.

26. **Cut Communication Lines:** Stop supplier attempts to subvert sourcing procedures by cutting off unauthorised communications

to influential personnel. Use workshops with internal stakeholders to reinforce the importance of maintaining integrity in sourcing and supplier relations. Shut off new business to a supplier that violates the policy.

27. **Fire Your Best People:** Remove your best team members from their day jobs and put them in charge of your most important Supply Management improvement initiatives. This ensures programme success and helps retain talent.

28. **Reward Good Behaviour:** When negotiating new business, be sure to give credit to preferred suppliers with good performance history. This will motivate them to continue to keep the good work.

29. **Empower Suppliers:** Use the web-based portal to empower suppliers to self-register and manage their profile, capabilities and certification information. This enhances buyer productivity, mitigates risks, helps prioritise tasks for supplier qualification development and rationalisation.

30. **Be Responsible:** Employ responsible supply strategies that foster fair labour practices, supplier development and diversity. Such practices assure supply often with a unique competitive advantage.

31. **Create Cross Experts' Team:** Hire Sourcing people with unconventional skills and experience. Those with the financial background can advance new costing models. Engineers can identify innovation; the technologist can speed successful deployments of supply management solution.

32. **Get Involved Early:** Involve buyer and commodity experts in the new product concept and prototype phases before the design is locked down. This allows supply managers to identify supplier innovations in the form of new technologies, manufacturing methods and processes that deliver quality cost and time to market benefits.

33. **Exploit Competition:** Benchmark pricing and supply trends to determine the right sourcing approach and to set realistic expectations for return prior to initiating negotiations. Ask for price improvement at least once in a year on various market-related context.

34. **Keep Your Winners:** Retain your top talents by using Competitive pay packages, keep your teamwork challenging, Create special projects for you top talent, develop training and mentoring programme, plot a clear career path for team members and support work-life balance.

35. **Hang Out With Losers:** Take time to debrief with suppliers that fail to win your business. The best will use your advice to develop new capabilities and bidding tactics that meet your future needs.

36. **Start Your Own News Network:** Use multiple channels to communicate the intent approach and results of Supply Management initiatives, i.e., newsletters, summarising sourcing projects, etc.

37. **Create a Crisis:** Whenever possible, link your Supply Management initiative to a top corporate goal or challenge, i.e., SOX compliance, China as marketplace, mitigating corporate risk are a good to start.

38. **Divide and Conquer:** Strategically segment your supply base and align sourcing and supplier management approaches by each segment. For instance, use more collaborative negotiations and align product and business roadmaps with your most strategic suppliers. Use competitive negotiations and low touch management methods for suppliers of commodity items with little strategic value.

39. **Become Multi-Lingual:** Gain support and respect from other groups by translating the benefits of your Supply Management initiatives into their language. For example, when sourcing advertising doesn't highlight cost savings, then tell marketing managers how you'll help ensure brand integrity, speed turnaround times and stretch their budget dollars.

40. **Ensure Currency:** Have supplier bid in your currency to reduce exposure to currency fluctuation or share currency risk with suppliers by negotiating tiered pricing in the seller's currency based on a mean price set at an agreed exchange rate.

41. **Choose Favourites:** Channel your most complex sourcing or contracting projects through a Centre of Excellence in which a core group of experts who are trained on the latest approaches and

functionality. This ensures the use of best practice and the delivery of faster and more consistent results. Empower less skilled users to specify projects and to execute common tasks such as basic e-RFS, reverse auction or configure contracts from pre-approved templates or terms.

42. **Define the Rules of the Game:** To guard against post-event bartering in supplier negotiations, use pre-requisites and knockout questions to ensure that suppliers agree to all conditions prior to the negotiation and to avoid any surprises or disputes after award.

43. **Make Sourcing a Team Sport:** There is a direct correlation between the involvement of non-purchasing stakeholders in the sourcing process and the contract compliance rates. Other functions can also offer in valuables insight into cost drivers and market dynamics.

44. **Get Aligned:** Ensure Sourcing and Operation management goals map directly to corporate and financial objectives.

45. **Use Your Own Papers:** Use your own standard form contracts whenever possible. This maintains consistency, reduces risks and improves negotiation leverage with suppliers.

46. **Close Backdoors:** To speed the success of new Supply or Contract Management initiatives, shut off legacy processes or system that could enable reluctant stakeholder or supplier to circumvent your new ones.

47. **Work Both Ends of the Organisation:** Deploying a Supply Management initiative requires securing support from top executives as well as frontline employees. Executives can provide the budget, resources and policy changes required to get a programme going. But success will be determined by stakeholder support and adoption.

48. **Come Armed with Facts:** Top executives want to know how your supply management programme is going to hit the bottom line and when it's going to get there. Ensure you have visibility into spending, compliance and performance date. Report how this intelligence links to corporate goals and profits.

49. **Make It Personal:** Ensure senior executives understand supply management contribution to your company profits, performance

and competitive edge by illustrating how supply management affects their personal lives.

50. **Put Risk Management on Autopilot:** reduce your exposure to supply risk by setting automatic alerts when key supplier milestones such as insurance renewals or critical certifications are coming to term. Issue reminders to supplier well in advance of the term date and track their compliance.

51. **Learn, Learn and Learn:** Enhance the skills of your supply management squad by partnering with the university. The top supply management schools offer executive training programmes, and most will customise the curriculum to meet your group needs.

52. **Get Classified:** Automate spend data classification to gain detailed insight into spending and to increase the accuracy and frequency of spend analysis. This can uncover hidden opportunities for better spend leverage, compliance and supplier rationalisation.

53. **Walk in Your Supplier's Shoes:** Prior to any negotiations, be sure to evaluate price and supply trends for key inputs into the product or services. This insight can help you identify key negotiation levers. It can also help you set realistic expectations for supply savings.

54. **Digitise Your Contract:** The number one challenge to spend compliance is poor visibility into contracts. This can improve contract compliance and minimise risk.

55. **Invest in Suppliers:** Dedicate key people to work with the main supplier to quickly assess the root cause of a problem and make recommendations on how to fix it. In most cases, the investment is less costly than switching to a new supplier.

56. **Start Off on the Right Foot:** Work closely with the new supplier to codify performance requirement measures and methods and continuous improvements goals for costs. Give supplier advance visibility into their performance linked development and new business opportunities.

57. **Encourage Counter Offers:** Enable supplier to bundle and un-bundle line items or provide alternative offers. This helps in determining the overall best value allocation. It allows suppliers to

differentiate on more than just price as suppliers are experts on their products, they may suggest innovative technologies or processes that you may never have considered.

58. **Go on a Diet:** Apply lean principles to sourcing and contracting, eliminate non-value added tasks such as redundant data entry, automate and streamline time and labour intensive processes and communications and embed standard policies, processes and controls in all aspects of supply management. Use repeatable sourcing and contract templates to enhance productivity and knowledge transfer.

59. **Know Your Market Share:** Measure spends from a corporate perspective, i.e., spend categories and know your market share with your key suppliers.

60. **It All Boils Down to Spend Under Management:** Spend under management is the percentage of company spends that is managed using approved strategic sourcing and compliance procedures.

61. **Add China Ginger:** Create a cost benchmark of the world's lowest one to sustain and retain the preferred supplier position.

62. **Set the Bar High:** Set annual goals for supplier cost reduction or innovation contribution. Make this goal part of supplier's overall performance scores.

63. **Should/Would/Could:** Uncover ways to offset increases in RM with saving in other areas like labour, logistics, etc.

64. **Use Explicit Language:** Clearly define the terms of sale for all your supplier contracts like fright, payment terms, insurance, etc.

65. **Don't Forget to Evolve:** The only thing worse than not measuring performance is tracking the wrong attributes. Revisit supplier performance metrics on a periodic basis to ensure that measures reflect changes in your company requirements and new supply market dynamics and risks.

66. **Take in a Show:** One of the quickest ways to catch up on the latest Supply and Contract Management best practices is to attend an industry association annual conference.

67. **Know Your Alternatives:** Identify alternative suppliers, substitution parts or materials at the outset of each new sourcing project. You may not award them any business, but it's good to have a record in case there is a problem with the primary supplier.

68. **Take It Off:** Work with internal stakeholders and suppliers to reduce packaging cost. This simple move not only cuts packaging costs but also can lower your inbound and outbound transportation and waste management bills.

69. **Avoid Late Fees:** Speed invoice matching and reconciliation to take advantage of those early payment discounts you spent so much time negotiating. Depending on their cash position, the supplier may be willing to offer additional rebates for earlier payments.

70. **Don't Neglect Your Soft Side:** Automating sourcing, contracting and procurement shortens process cycles by 50% to 70%. Cycle time improvement free you up to apply strategic sourcing to a wider range of spend.

71. **Reap What You Sow:** Monitor transactional and pricing compliance with the contract. Report compliance performance on a periodic basis – fingering areas of non-compliance.

72. **Make Connections:** When sourcing from the low-cost region, foster frequent interaction between supplier's quality personnel and your engineer's. This helps ensure quality and can speed the ramp-up process.

73. **Mix Centralised Decision with Local:** When embarking on a supply or contract management improvement initiatives, strive for centrally defined processes and global spend leverage with geographically distributed support and execution.

74. **Give Suppliers More Responsibility:** Reduce costs and exposure by adopting alternative business relationship with your top performing suppliers or for suppliers in categories where you lack expertise.

75. **Make Plans:** Define a 3-year roadmap for transforming your Supply Management organisation and performance. Be sure to include measurable milestones as well as resources and policy changes required to achieve them.

76. **Don't Be Afraid of Commitment:** Assess future pricing and supply market implications to determine when to engage in long-term supplier or hedging agreements. This will assure supply and lock in preferred pricing, which is particularly wise in the face of rising metal, plastics and energy prices.

77. **Play a Fair Game:** Clearly define requirements and rules of business including criteria for supplier selection. Be sure to abide by your own stated rules and if you need to make any change share it equally with all suppliers. Doing so will ease suppliers concerns and maximise their competitiveness.

78. **Don't Recreate the Wheel:** Leverage the know-how of your solution provider to learn the method for effective sourcing and Contract and Supply Management, many times suppliers have information critical for success. This helps in avoiding the mistakes made by early adopters.

79. **Dilute Conflicting Situations:** Start your Supply Management transformation in high impact and low friction areas to gain acceptability. Parameters like strategic sourcing that don't require multiple functions and drive measurable savings. This will allow in demonstrating the value of your programme and funds for additional improvement initiatives.

80. **There Is No Substitute for F2F Meeting:** Supplier management is a continuous process to bridge the gap between your organisation supply goals and market condition. Ensure to have regular scheduled meetings with your functional and business unit as well as supplier base so that supply goals can be aligned with corporate objectives.

81. **Honour Top Suppliers:** There is strong evidence that publicly recognised top performing suppliers encourage performance improvement across the supply base. Showcase them as the preferred solution.

82. **Transparency Is the Key:** Manage internal expectations by clearly communicating sourcing or contracting programme and expected results. Issue project sheet for each deal showing negotiation approaches, pricing and market trend and predictions for results.

83. **4 Is the Magic Number:** To exploit competition and maximise cost advantage includes at least four suppliers in every negotiation.

84. **Pay to Stay:** There is a direct correlation between Supply Management team salaries and the performance and value returned from sourcing. Industry benchmark report top Supply Management organisation pay 30% higher salaries and outperform peers by more than 2X in-process efficiencies, effectiveness and cost reductions.

85. **Promote Six Sigma Culture:** Using six sigma techniques in daily business activities will dilute error and enhance the accuracy of outcome as per plan.

86. **Talk Co-Operative, Behave Competitively:** Supplier should look at you as a co-operative alliance, whereas you need to groom them to become a globally competitive source.

87. **Visit Your Supplier Facilities:** Visiting supplier facility will update on supplier capability aspect as well as culture and best practices used at plant level for a sustainable relationship.

88. **QA Lab. Is Cockpit of Plant:** While visiting plants/factories, minutely observe things at QA laboratory of the supplier. It is said that if the laboratory is not in order likelihood of maintaining plant housekeeping is very low.

89. **Ethics Are Backbone:** Show a high ethical standard while dealing with your suppliers; it preserves respect for you as well as the goodwill of your employer.

90. **Career Development:** Offer career to your team not only the job or work, but this can also be done through job-rotation and will develop internal grooming process on the job for the growth of individuals to take up increased responsibilities within the organisation.

91. **Business Alignment Meetings (BAM):** Ensure for proactively planned meetings with key suppliers, develop monthly calendar in context to whom to meet and what to discuss (i.e., well-defined agenda).

92. **Achieve Greater Outcomes:** Improved strategic negotiations and Supplier Relationship Management can drive a positive, sustainable outcome.

93. **Leverage System and Processes:** To reduce price and drive cost avoidance.

94. **Create Improved Communication:** Effective and strategic communication with valuable clients/suppliers will help in improving bottom line negotiations.

95. **Drive Effectiveness and Efficiencies:** Do proactive procurement strategies and negotiations instead of a reactive action plan.

96. **Employ Proven Methodologies:** It helps in managing procurement and negotiation challenges, so following steps, methods, templates, white paper and other experience make sense.

97. **Minimise Risk:** Proactive planning, risk management and negotiations with supplier network will help in minimising supply-chain disruption and risk involved.

98. **Quantify Best Practices:** Strategic negotiation and procurement processes, best practice and system, templates, form, contacts can provide sustainable results.

99. **Establish Benefits:** Utilise strategic procurement outsourcing and offshoring to organisation benefit.

100. **Host Lunch to Your Key Supplier's at least Once in Six Months:** Take your key suppliers for lunch at least once in 6 months and talk about prices on a one to one basis post your food. With that gesture, supplier thinks you are a good guy and typically build a personal rapport and offers you a reasonable price compared to their other customer.

Bibliography

1. Arzu Tektas & Aycan Aytekin (2011) ~ Supplier selection in the International Environment: A Comparative case of a Turkish and an Australian Company—IBIMA Publishing.

2. Chris M. Wilson and Catherine Waddams (2006). ~ Do consumers switch to the best supplier?–School of Management, University of East Anglia.

3. Childe S.J. (1998) ~ The extended concept of Cooperation–Production Planning & Control–The Management of Operations.

4. Dickson. G.W. (1966) ~ An analysis of Vendor selection systems and decisions–Journal of Purchasing.

5. Ellram, LM ~ Supplier selection decision in Strategic partnership–Journal of purchasing and materials management. Vol. 26

6. Humphreys, PK, Wong Y.K., Chan F.T.S. (2003) ~ Integrating environmental criteria into supplier selection process–Journal of materials processing technology.

7. Hendricks, K.B., V.R. Singhal. 2003b. "An Empirical Analysis of the Effect of Supply Chain Disruptions on Operating Performance." Working Paper, DuPree College of Management, Georgia Institute of Technology, Atlanta, GA.

8. ICCA–Contribution of the international council of chemicals associations (ICCA)–Journal of International Council of Chemicals Associations (2001).

9. Kraljic, P (1983) ~ Purchasing must become supply management–Harvard Business Review. 61.

10. Luitzen De Boer, Eva Labro, Pierangela Morlacchi (2001) - Review of methods supporting supplier selection–European Journal of Purchasing & Supply Management. Vol. 7.

11. Motwani, J, Youssef M, Kathawala Y, Futch E (1999) - Supplier selection in developing countries: A model development–Integrated manufacturing systems.

12. Prahalad CK and Hamel G (1994)–Strategy as field of study–Why search for a new paradigm–Strategic Management Journal.

13. Suresh Babu, T. & Kamana, S. (2005) - An Analytical Hierarchy process for vendor election evaluation–South Asian journal of management. 12.

14. Sim HK and Mohamed K. Omar (2010) - Survey on supplier selection criteria in the manufacturing industry in Malaysia.

15. Till Vestering, Ted Rouse, Uwe Reinert and Suvir Varma (2005) - Making the move to low-cost countries–BAIN & Company.

16. Verma R. and Pullman ME (1998) - An analysis of the supplier selection process–Omega International journal of management science.

17. Vonderembse, Mark A & Michael Tracey (1999) - Impact of supplier selection criteria and supplier involvement in manufacturing performance–Journal of Supply chain Management.

18. Wilson EJ (1994) - Relative importance of supplier selection criteria: A review and update–International journal of purchasing and materials management.

19. Weber, CA Current, JR & Benton, WC (1991) - Vendor Selection Criteria and methods–European journal of operational research.

20. Wadhwa V. Ravindan A (2007) - Vendor Selection in Outsourcing, "Computer & Operations Research".

About the Author

Aditya Verma has been associated with various segments of industries in Indian/global business, i.e. FMCG, chemicals & polymers, food and beverages, oil and gas, electronics and home appliances, etc. He holds a Ph.D. in Supply Chain/Supplier Management and also has post-graduate level degrees and diplomas in the streams of operations management, materials management, marketing and sales management, international trade, export/import management, packaging technology & healthcare and nutrition and is a certified Six Sigma Green Belt holder as well.

He has published many papers and articles in leading business magazines largely with respect to procurement and materials management. Some of the articles have covered topics such as,

- Empowering Indian speciality chemical companies to be competitive in global sourcing/supply chain.

- Buyer's perspective on global outsourcing of their chemical supply chain.

- Select Green while choosing New Supply Source.

- How cultural norms and values affect the negotiation process.

He worked for four major multinational companies in the FMCG & Chemical World, where he owned the prime responsibility of developing India as a supply base for their global manufacturing units from ground zero. He later created a multimillion-dollar sourcing base, which actually inspired him to write this book and share the knowledge and experience with the younger generation w.r.t. practical/applied aspect. His versatile experience in global multinational companies and Indian – home-grown domestic companies created a fusion of "GloCal Sourcing", i.e. Global + Local. He loves to share his procurement and supply-chain related experience with the younger generation, who plan to take Supply-Chain or Purchasing as their career option.

He graduated through his experience. He started his career almost 28 years back as a Purchase Trainee. Learning the fineness of a function in this competitive business battlefield with companies across the globe, learning cultural aspects and picking up best of "Cross-global" culture and imparting the same during professional tenure, many of the people who worked with him in the past have gained so much knowledge and grown so well in their career, and are now heading the procurement and sourcing function in many multinational companies at regional and global level. His inspiring leadership motivated many youngsters to change their work stream to procurement.

This book is an attempt to share that learning and knowledge with a larger population either in business and want to understand supply-chain market dynamics for their organisation or are into supply-chain profession or if they are working in Sales and Marketing function and want to understand the psychology of other side of the table. This book will extend help in forming the right strategy.

Aditya's professional mantra is "Procurement is the foundation and it impacts P&L directly, every single $ saved in procurement adds 100% to the company bottom line or profitability," therefore skilled procurement person can make a huge difference to organisation profitability. This has been the prime intent of the author in conceiving this book "**Supplier Matters.**"

www.ingramcontent.com/pod-product-compliance
Lightning Source LLC
Chambersburg PA
CBHW021353210526
45463CB00001B/91